Interweaving Equitable Participation and Deep Mathematics

Building Community in the Elementary Classroom

Susan Jo Russell

and

Deborah Schifter

with

Collaborating Teachers: Quayisha Clarke, Emmanuel Fairley-Pittman, Natasha Gordon, Jeff Parks, Isabel Schooler, and Michelle Sirois

and

Critical Friends: Cynthia Ballenger, Virginia Bastable, Yi Law Chan, Marta Garcia, Lynne Godfrey, Hetal Patel, and Darlene Ratliff

CORWIN Mathematics

FOR INFORMATION

Corwin

A SAGE Company

2455 Teller Road

Thousand Oaks, California 91320

(800) 233-9936

www.corwin.com

SAGE Publications Ltd.

1 Oliver's Yard

55 City Road

London EC1Y 1SP

United Kingdom

SAGE Publications India Pvt. Ltd.

Unit No 323-333, Third Floor, F-Block

International Trade Tower Nehru Place

New Delhi 110 019

India

SAGE Publications Asia-Pacific Pte. Ltd.

18 Cross Street #10-10/11/12

China Square Central

Singapore 048423

Vice President and
   Editorial Director: Monica Eckman

Associate Director and
   Publisher, STEM: Erin Null

Senior Editorial Assistant: Nyle De Leon

Production Editor: Tori Mirsadjadi

Copy Editor: Michelle Ponce

Typesetter: C&M Digitals (P) Ltd.

Proofreader: Barbara Coster

Indexer: Integra

Cover Designer: Scott Van Atta

Marketing Manager: Margaret O'Connor

Printed in the United States of America

LCCN 2024028862

This book is printed on acid-free paper.

24 25 26 27 28 10 9 8 7 6 5 4 3 2 1

WHAT YOUR COLLEAGUES ARE SAYING . . .

"While reading this book and watching the videos, I found myself wishing that I had read it when I was a classroom teacher. The authors give readers a window into a learning community of educators as they apply their new sense-making about mathematics to their teaching. While the academic content is mathematics, we are guided through the creation of a community where each learner's sense of self and agency is strengthened around their own learning."

Janice E. Jackson
Former Deputy Superintendent, Boston Public Schools
Former Deputy Assistant Secretary, Office of Elementary and
Secondary Education, U.S. Department of Education Newark, CA

"This book is a must-read for math educators across K–12 who want to foster curiosity, meaning making, and collective mathematical agency in our nation's diverse classrooms. The authors and their teacher-leader partners provide multiple resources that illustrate equitable community-focused instruction where students take charge of and practice essential mathematical skills including representing, conjecturing, and generalizing. Powerful!"

Julia Maria Aguirre
Professor of Education, University of Washington Tacoma
Tacoma, WA

"This resource provides images of teaching that promote student voice and agency in mathematics classrooms. The authors' creative approach—with video clips, transcripts, and commentaries from collaborating teachers and critical friends—will engage teachers, coaches, and leaders across the professional continuum. The images, mathematical representations, questions, and activities will transform your thinking and practice and ultimately elevate student participation and learning."

Kathryn B. Chval
Dean and Professor, Mathematics Education, University of Illinois Chicago
Chicago, IL

"I sum up *Interweaving Equitable Participation and Deep Mathematics* as: Deep Mathematics Content + Community Participation that leads toward equitable engagement, growth, and outcomes. Teachers and leaders can engage in deep professional learning throughout the sections in this book and will reflect on their practices and students' learning."

Robert Q. Berry, III
Dean, College of Education, The University of Arizona
Tucson, AZ

"Central to this book is the valuing and foregrounding of children's mathematical thinking. Through classroom videos and supporting reflection prompts and activities, the authors show the interplay of providing rich mathematical experiences and focusing on equitable participation. This results in a powerful resource for teachers, teacher educators, and anyone interested in a mathematics education that honors all children's ideas."

Marta Civil
Professor and Roy F. Graesser Chair, The University of Arizona
Tucson, AZ

"In this exciting book, Russell and Schifter invite us to join their professional learning community to explore teaching that interweaves a commitment to equity and rigorous mathematics. Each chapter offers vivid examples of teachers and students engaged in rich mathematical tasks and deep collaborative conversations in their classrooms. It really is a privilege to reflect alongside these authors. I can't wait to share this book with my teacher education students."

Sandra Crespo
Professor and Associate Chair of Graduate Education,
Department of Teacher Education,
Michigan State University
East Lansing, MI

"Together with collaborating teachers and critical friends, Russell and Schifter take us into inspirational and real classrooms, revealing the complexities and insights that come with taking students' intellects seriously and nurturing equitable classroom communities. This book invites rich reflection and entices you to move with urgency and eagerness to develop your mathematics teaching and make real your commitments to racial equity. You will return to it again and again."

Elham Kazemi
Professor, Mathematics Education, University of Washington
Seattle, WA

"Russell and Schifter have created a remarkable resource for teachers to gather around! The authors have expertly combined rich videos with insightful analysis and thought-provoking questions. Readers will have the best sort of collaborative, empowering opportunities to grow both a deep understanding of mathematics teaching and learning and their capacity to provide equitable instruction to all students. This book is a gift!"

Tracy Johnston Zager
Math Coach
Author, *Becoming the Math Teacher You Wish You'd Had*
Coauthor, the *Building Fact Fluency* toolkits
Portland, ME

Contents

Part Three: Student-Created Representations Offer Anchors, Openings, and Depth

Part Three: Student-Created Representations Offer Anchors, Openings, and Depth 127

Part Four: Students Are Initiators and Advocates for Their Own Learning

Part Four: Students Are Initiators and Advocates for Their Own Learning 183

Visit the companion website at
companion.corwin.com/courses/equitable-deep-math
for downloadable resources.

Note From the Publisher: The authors have provided video and web content throughout the book that is available to you through QR (quick response) codes. To read a QR code, you must have a smartphone or tablet with a camera. We recommend that you download a QR code reader app that is made specifically for your phone or tablet brand.

Videos may also be accessed at
companion.corwin.com/courses/equitable-deep-math

Preface

The idea that sparked the project that eventually resulted in this book came from our colleague, Elizabeth Sweeney. "You have to capture these teachers and their students on video," she told us one summer afternoon over lunch by the seashore. By "these teachers," she meant graduates of the Master's program she worked in who were now teaching in public, neighborhood schools in Boston, Massachusetts. Liz knew we had often filmed classroom examples in previous projects, including in her own classroom twenty years earlier when she was a fifth-grade teacher. She believed, as we do, that, when the focus is on students' sense-making, video can be an important tool to slow down and show what learning and teaching can look like—with real students and real teachers in real time. Liz was now mentoring teachers who had a strong commitment to equitable participation in schools with significant populations of Black and brown students—groups that have been historically marginalized in mathematics (Joseph & Alston, 2018; Ladson-Billings, 2006, 2007). Liz—and we—saw an opportunity for a partnership in which a program of rigorous mathematics would be intertwined with a focus on supporting students' identity and agency as mathematics learners in a classroom mathematics community. We and the teachers would learn together, analyzing video and written documentation of the students' work. We believed from the beginning that this collaboration would lead to rich and thought-provoking professional development resources.

We first came to know Liz, a long-time employee of Boston Public Schools, as a teacher-participant in a professional development program we were leading. Reflecting on her view of mathematics teaching when she entered the program, Liz said, "I never had the belief that math was about thinking, reasoning, or making sense. It was just learning the rules. There was no *self* in math." Once Liz began to see mathematics as a realm of exploration, she discovered she loved the changes her growing engagement with mathematics brought about in her preparation, her teaching, and her classroom, and she saw her fifth graders respond enthusiastically as she engaged them in thinking deeply about mathematical ideas. Liz taught fifth grade for many years, then went on to become a math coach and, later, a citywide leader of elementary mathematics. Upon retirement from Boston Public Schools, Liz joined the faculty of the Boston Teacher Residency (BTR).

BTR is a joint initiative of the Boston Plan for Excellence and Boston Public Schools that combines a year of residency in a school setting working with an experienced teacher and targeted master's-level coursework. As BTR math faculty, Liz provided experiences like those in her own professional development that had changed her beliefs and practices about mathematics and mathematics teaching. During their residency, the teachers she was supporting engaged in doing mathematics for themselves and were learning how to establish a classroom community based on students sharing their thinking and building concepts together about important mathematical ideas. Once they began teaching in their own classrooms, Liz continued as their mentor through BTR. Following up on Liz's suggestion, as plans for a joint project developed, we were able to obtain support both from BTR and from TERC, a nonprofit organization in Cambridge, Massachusetts, that focuses on mathematics and science education.

Liz proposed that the collaborating teachers would use lesson sequences that we had recently published (Russell et al., 2017). These eight lesson sequences investigate sets of related generalizations about the basic operations: addition, subtraction, multiplication, and division. Written for Grades 1 to 5, each sequence consists of about twenty 20-minute sessions. The sequences are designed to be used in addition to the regular math program, much like Number Talks or other such routines.

The videos at the heart of this book come from the classrooms of six teachers, graduates of BTR, who were teaching from these lesson sequences. For two years, we visited the teachers' schools, video recorded lessons, met with the teachers individually and as a group, and exchanged writing about what we and they saw in the classroom. Susan Jo was behind the camera, while Deborah kept a running record of the class session. You'll notice Deborah, and sometimes also Liz, taking notes in some of the videos. After the recording portion of the project was complete, we asked each teacher to review the videos from their classroom and interviewed them about their decisions, their questions, and what they learned. We also selected video clips to discuss with our own study groups and to present at workshops and conferences.

Here is a little information about each of the Collaborating Teachers you will be meeting in this book:

Quayisha Clarke is an experienced educator with a decade of teaching within the Boston Public School system. She holds National Board Certification and currently serves as an Instructional Coach at Dudley Street Neighborhood Charter School. In addition to her professional commitments, she is dedicated to furthering her knowledge and is pursuing a PhD specializing in Human Development and Learning at Lesley University. With her wealth of experience and ongoing academic pursuits, Ms. Clarke is committed to delivering high-quality education and support to students, teachers, and educational institutions. Ms. Clarke was a Grade 2 teacher during the project.

Emmanuel Fairley-Pittman is an Inclusive Education Coach supporting teachers and schools in Boston to plan for and implement inclusive practices. As a National Board certified educator having taught third, fourth, and fifth grade at the Grew School in Hyde Park for eight years before this role, he is committed to providing access for all students so that they have meaningful academic and social experiences in school. He attributes his commitment to the equitable teaching of mathematics in part to his own teachers growing up, specifically Ms. Langston and Mrs. Hunter. These teachers believed in his abilities and provided the necessary tools and support for him to grow into his identity as a mathematician. Mr. Fairley-Pittman was a Grades 3–4 teacher during the project.

Natasha Gordon's collaboration with exceptional math coaches enabled her to cultivate curiosity and foster independence in her first-grade students. Her passion lies in ensuring equal opportunities for all learners, fostering success, and embracing diverse ideas. Currently serving as an equitable literacy coach, she strives to continue to create and uplift inclusive educational environments. Ms. Gordon was a Grade 1 teacher during the project.

Jeff Parks is a facilitator of professional learning for the Telescope Network and math instructional coach at the Mather Elementary School. Before this, Mr. Parks taught third grade in Dorchester at the Mather Elementary and Everett Elementary for a combined 10 years. He has served in these schools as the Boston Union Rep, teacher leader, and tech coordinator. He has also worked with Boston Teacher Residency as a graduate coach and facilitator. His commitment to group work and collaborative problem solving in mathematics continues to reinforce his belief that all students can do math at the highest level. Mr. Parks was a Grade 3 teacher during the project. He will be returning to the classroom next year and is excited to teach math again with his new fourth-grade students.

Isabel Schooler taught first grade for six years, then worked as an elementary special education teacher for a year. In her current role as a math interventionist and coach, she draws on her experience with outstanding math coaches, finds inspiration in the thinking of young mathematicians, and remains dedicated to equity in mathematics. She looks forward to returning to the classroom this fall! Ms. Schooler was a Grade 1 teacher during the project.

Michelle Sirois has been teaching for eleven years. After graduating from the Boston Teacher Residency Program, she taught in Boston Public Schools for eight years. She currently is a Math Specialist for Grades 3–5 in Milford Public Schools. Throughout her years in the classroom, she has focused on building communities of mathematical learners where students notice patterns, ask questions, and play with numbers and ideas. She truly believes that children are sense-makers. Ms. Sirois was a Grade 4 teacher during the project.

As we, the authors, began to tease out the interwoven themes to be addressed in this book, we sought out the perspectives of other educators to view video from the classrooms of these six Collaborating Teachers through the lenses of their different experiences and backgrounds. We are both native English speakers of European descent, and we wanted to be informed by other educators from a variety of backgrounds who, throughout their careers, have focused on issues of identity, race, language, and inclusion. These educators, whom we call our "Critical Friends," provide analysis and raise their own questions throughout the book. Below is a quick introduction to each of our Critical Friends:

Cynthia (Cindy) Ballenger received a doctorate in Applied Linguistics from Boston University. She has worked as a teacher in public schools for many years and as a professor and program director at Tufts University. A central focus of her teaching has been the role of talk and culture in children's learning and participation, as in her books *Teaching Other People's Children: Literacy and Learning in a Bilingual Classroom* and *Puzzling Moments, Teachable Moments: Practicing Teacher Research in Urban Classrooms*.

Virginia Bastable is one of the authors of the *Developing Mathematical Ideas* professional development curriculum published by NCTM and two books on mathematical reasoning about the operations: *Connecting Arithmetic to Algebra* and *But Why Does It Work: Mathematical Argument in the Elementary Classroom*. She also contributed to the third edition of *Investigations in Number, Data and Space*. Before retiring, Dr. Bastable was the Director of the Mathematics Leadership Program and its precursor, SummerMath for Teachers, at Mt. Holyoke College. She started her education career as a high school math teacher, a job she enjoyed for more than twenty years. Her current work includes facilitating online academic year courses and summer institutes for the Master of Arts in Teaching Mathematics at Mount Holyoke.

Yi Law Chan is a New York City-based school leader where she is prioritizing social-emotional wellness and student-centered instruction. She brings to her current work over twenty years of experience as a former classroom teacher, math coach, assistant principal, and math specialist. In these roles, she organized and facilitated professional learning communities in examining the impact of equity-based teaching practices and teacher content knowledge on student learning.

Marta Garcia is an elementary mathematics specialist/coach with over thirty years of experience in teaching, coaching, and facilitating professional learning. She currently works as a mathematics coach with a variety of school districts across the U.S., teaches graduate courses focusing on mathematics leadership, and is the co-host of a virtual professional network of math coaches. She received the Presidential Award for Excellence in Mathematics and Science and the

NCCTM Rankin Award for distinguished service in the area of mathematics teaching and learning. Much of her current work focuses on how to empower teachers in the development of equitable classrooms where the voices of students, including multilingual learners and those from marginalized groups, are lifted.

Lynne Godfrey designs and facilitates professional development with coaches, teachers, and administrators to develop and sustain ambitious, equitable learning communities in their schools. As an educator, she served as a classroom teacher of Grades 2–8, a math coach in the Boston Teacher Residency program and elsewhere, and a director of instruction. Her work with Bob Moses and the Algebra Project for over thirty years, both locally and nationally, has influenced and sustained her ongoing commitment to access and equity for all adults and children in mathematics.

Hetal Patel is a mathematics instructional leader in New York City where she facilitates professional learning for teachers and school-building leaders. Additionally, she was an assistant principal, a school-based mathematics coach, and a math intervention teacher. In prior years, she facilitated several of the *Developing Mathematical Ideas* modules at Mount Holyoke College and nationally. Her professional interests include using Japanese Lesson Study to build professional learning communities, learning more about equity and developing identity in math education, and cultivating curiosity and joy in mathematics teaching and learning.

Darlene Ratliff, a recent retiree from Boston Public Schools, worked with general (K–8) and special education populations as an administrator, special education teacher, math coach, and math facilitator. She has served on the Assessment Development Committee for the Massachusetts Department of Elementary and Secondary Education, taught at the Horace Mann School for the Deaf and Hard of Hearing, and is a nationally certified American Sign Language interpreter. She has facilitated professional development workshops locally and nationwide, including *Developing Mathematical Ideas* seminars and *Looking at Student Work* sessions. Through witnessing firsthand the productivity and potential of all students, she developed her mantra: "Regardless of differing abilities, all students have the potential to achieve."

As you read this book and view the classroom videos, we invite you to be in conversation with many voices—with us, with the Collaborating Teachers, with our Critical Friends, with the many students you will observe shaping their mathematics identities—and, for the best experience in using this resource, to reflect about your own students with your own colleagues. Listening to and respecting all of these voices provides openings for honesty and persistence in efforts to create classroom communities that interweave rigorous mathematics and equitable participation.

Susan Jo Russell
Deborah Schifter
June 2024

Acknowledgments

Many people have contributed to the project that produced this book. Without the passion and insight brought to their teaching by the Collaborating Teachers, their willingness to have their lessons videotaped and to meet with us to reflect on students' work, and their writing included here, this book would not exist. We cannot thank enough Quayisha Clarke, Emmanuel Fairley-Pittman, Natasha Gordon, Jeff Parks, Isabel Schooler, and Michelle Sirois. Similarly, we are so grateful to our partners in reflection, our "Critical Friends"—Cindy Ballenger, Virginia Bastable, Yi Law Chan, Marta Garcia, Lynne Godfrey, Hetal Patel, and Darlene Ratliff—who met with us, viewed and reflected on the classroom videos, and spoke from their hearts with the intensity and determination they bring to all of their work. Short bios of the Collaborating Teachers and Critical Friends can be found in the Preface. We want to thank Liz Sweeney, who started us on the path that led to this book (see the Preface for that story), the Boston Teacher Residency for making Liz's involvement with the project possible, and particularly Julie Sloan, then Director of the BTR Early Career Teaching Network. Christine Connolly, a Boston Public Schools principal at the time of this project, worked with us to implement professional development for the teachers and was instrumental in arranging our visits, videotaping, and meetings with the teachers. We are grateful for the support of Linda Davenport, then Boston Public Schools Director of K–12 Mathematics and a long-time colleague, for her support as our collaborating department head for Boston Public Schools. And, throughout the project, members of the Professional Development Study Group, a cross-institutional forum in the Boston area that has met for many decades to share work in progress, provided critique and support as we shared video clips from our work. We thank the Education Research Collaborative at TERC, TERC President Laurie Brennan, and the TERC Board of Trustees for recognizing the potential of this collaborative opportunity and funding this project. And we are grateful to the associate editor and publisher at Corwin, Erin Null, for her belief in and support of this book.

Finally, this book would not exist without the many students in the classes we observed and videotaped. We learned from them, were surprised by them, and were, again and again, struck by the joy and seriousness with which they pursued their ideas, once given the opportunity to voice them. We are delighted to be able to share their work with readers of this book.

Publisher's Acknowledgments

Corwin gratefully acknowledges the contributions of the following reviewers:

Michelle D. McKnight
Mathematical Instructional Coach, South Windsor Public Schools
Manchester, CT

Georgina Rivera
Principal, Charter Oak International Academy
Hartford, CT

About the Authors

Dr. Susan Jo Russell began her career in education as a K–3 classroom teacher and elementary mathematics coach. For the last four decades, she has been a senior researcher at TERC, a nonprofit organization that focuses on mathematics and science education, where she directed projects on children's mathematical understanding and on supporting teachers to learn more about mathematics and about children's mathematical thinking. She spearheaded the original development and second edition of the K–5 mathematics curriculum, *Investigations in Number, Data and Space,* and contributed to the launching of the Forum for Equity in Elementary Mathematics. In recent years, her research has centered on how students engage in early algebra, number and operations, and mathematical argument, and how to support teachers to engage all of their students in high-level mathematical reasoning.

Dr. Deborah Schifter has worked as an applied mathematician; has taught elementary, secondary, and college level mathematics; and, since 1985, has been a mathematics teacher educator and educational researcher at Mount Holyoke College and at the Education Development Center. She authored *Reconstructing Mathematics Education: Stories of Teachers Meeting the Challenge of Reform* and edited a two-volume anthology of teachers' writing, *What's Happening in Math*

Class? for which she received the American Educational Research Association Professional Service Award in recognition of an outstanding contribution relating research to practice. She was a writer for *The Mathematical Education of Teachers* as well as the second and third editions of the K–5 curriculum, *Investigations in Number, Data, and Space.* Her recent research has focused on students' mathematical generalizations and how students use a variety of representations to explain why such generalizations are true.

Deborah Schifter and **Susan Jo Russell** have collaborated on many projects. With Virginia Bastable and a group of collaborating teachers, they produced the professional development series, *Developing Mathematical Ideas,* which is designed to help teachers think through the major ideas of K–8 mathematics and examine how children develop those ideas. With Virginia and another group of teachers, they developed *Connecting Arithmetic to Algebra*, which shows how investigating the behavior of the operations can move K–6 students forward. *But Why Does It Work? Mathematical Argument in the Elementary Grades*, coauthored by Susan Jo, Deborah, and others, is a resource for teachers who want to learn how to integrate mathematical argument into their instruction. (See the Resources section for further information.)

Introduction

What Is a Mathematics Community?

What does a mathematics community look like in an elementary classroom? How do we—teachers, coaches, administrators, all of us who support student learning—engage young mathematicians in deep and challenging mathematical content? How do we ensure that every student contributes a voice to this community, including students who have been historically marginalized in mathematics, students who have not believed they have mathematical ideas that are important to share, or who, when they have tried to express their ideas, have not been heard? These are the core questions this book seeks to address.

This book focuses on the interweaving of two commitments to children: a commitment to teaching deep and challenging mathematics and a commitment to equitable participation for all students in the classroom community. Without the opportunity for students to engage in significant mathematical content, a focus on equity is empty. If there are systems in place to ensure that every student speaks, but the math content is superficial and devoid of sense-making, we are not preparing students to become mathematical thinkers. On the other hand, if we attend exclusively to the rigor and depth of the mathematics, a few students may dominate, and what's perceived to be a correct and complete response from one or two students may stop continued discourse. Without attention to how each student engages with the content, the depth of the mathematics makes no difference for too many students.

In the intersection of deep mathematics and equitable participation, we have classrooms in which the *mathematics* content is significant, and the *community* enables each student to grow in understanding through their participation. In these classrooms, each student is assumed to have mathematical ideas, and it's the work of all of us to learn to listen for them. But as classroom teachers, coaches, instructional leaders, and others responsible for students' learning, how do we build and sustain such a community?

Four Aspects of a Mathematics Community

For two years, we, the authors of this book, undertook a research project in which we visited and videotaped lessons in the classrooms of six public elementary school teachers who were working to create classroom communities in which all students were engaged in serious content. Throughout this book we will refer to them as our Collaborating Teachers. We documented lessons, collected student work and teachers' writing, and reflected on these lessons with the teachers in order to uncover key ideas that characterized their evolving mathematics communities. As you move through the book, you yourself will interact with the videos and teachers' writing, observing and reflecting on students' thinking as well as the actions and thoughts of the teachers as they build mathematical communities in their classrooms. You will also hear the reflections and observations of our Critical Friends, who brought their own experiences to bear on what they saw in these classrooms. We'll explain more about this in a bit.

For our Collaborating Teachers, community was characterized by four main ideas: *every voice matters; collaboration supports student agency; student-created representations offer anchors, openings, and depth;* and *students are initiators and advocates for their own learning.* This book is therefore structured in four parts, each focusing on issues related to one of these ideas. While these four aspects of community interact and strengthen each other, we separate them in order to dig more deeply into what it takes to build each facet of a mathematics community (see Figure Intro.1).

Part One: Every voice matters.

In a mathematics community focused on student thinking, teachers establish classrooms in which students learn to take on the responsibility of sharing their ideas and attending to those of their classmates. The chapters in Part One focus on the importance of trusting students to take on challenging ideas as they develop their *identities* as doers of mathematics—how they view their own confidence and competence in approaching mathematical problems and questions. The videos illustrate how students are offered many modes of participation as they are beginning to develop their *mathematical agency*, that is, their inclination and ability to rely on their own reasoning. Both students who are eager to contribute their ideas and students who need more time and support to articulate their thoughts are included in the classroom discourse. Teachers and students listen intently to discern the sense in each student's thinking.

The culture of a mathematics community, as seen in the videos, is established through the teachers' sustained and intentional efforts. In Part One, our collaborating teachers discuss the norms they set and the tools they provide in the beginning of the year to help students understand what it means to

Interweaving Equitable Participation and Deep Mathematics

Figure Intro.1 • Four Aspects of a Mathematics Community

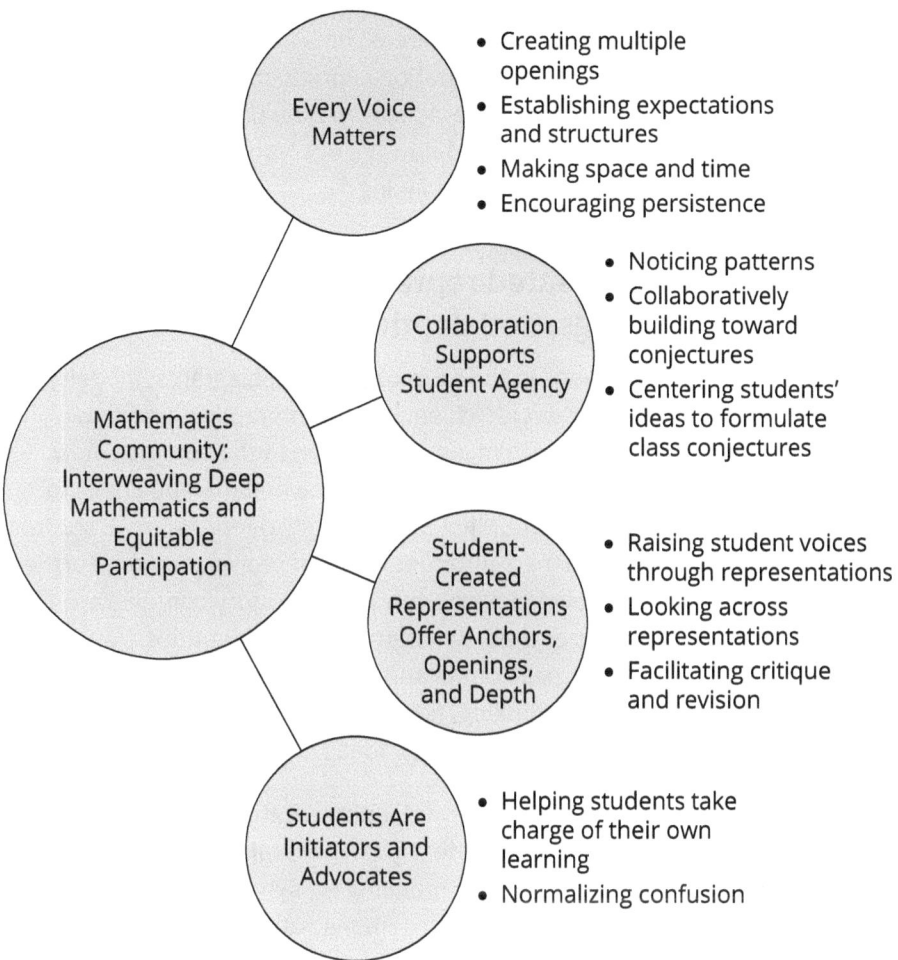

contribute to a mathematics discussion. Maintaining that culture requires the efforts of all participants—students and teacher—throughout the year.

Part Two: Collaboration supports student agency.

Mathematics is about complex ideas. Even in the primary grades, if given the opportunity and the tools, students engage in deep and abstract mathematical ideas. Unlike the image of the solitary, brilliant student working alone to solve every problem, much of mathematics requires collaboration, or what Aguirre et al. (2024) call "collective mathematical agency": "Classrooms of students can exhibit *collective mathematical agency* when teachers and their students act together to solve problems, working from the shared belief that viable strategies can be developed and solutions can be found. Different students can contribute different elements to this collective agency" (p. 17).

The chapters in Part Two focus on how students build ideas together as they notice patterns and articulate conjectures based on those patterns. The videos and commentary illustrate several related ideas: how offering partially formed or not yet well-articulated ideas for consideration is productive and, in fact, critical to the work of the community; how the practices of questioning and revising contribute to the community's ideas; and how it's everyone's responsibility to try to understand and build on each other's thinking.

Part Three: Student-created representations offer anchors, openings, and depth.

Student-created representations in the form of pictures, diagrams, models, and story contexts ground student ideas, encourage interaction, and draw more students into mathematical thinking. In contrast to the common view that students must be weaned off the use of diagrams and manipulatives to engage in abstract realms, understanding mathematics deepens for all students when they make connections across different forms of representation. Further, creating and explaining their own pictures, models, and story contexts are key parts of students' expressions of their own mathematical identities. You will see in Part Three how students are passionate and engaged as they use their own representations to explain their thinking, field questions from other students, and revise their representations to make their ideas clearer.

The chapters in Part Three present students' representations of pairs of related equations or story problems that illustrate a given generalization. By referring to their representations, students create meaning for symbols. Representations can also be key to mathematical argument, demonstrating *why* a procedure works or *why* a generalization is true.

The videos show students working together to interpret a set of representations chosen by the teacher from those created by the class. Commentary considers such issues as the connections made by looking at different representations, how a teacher selects which representations to share, and the value of choosing some representations that may need revision.

Part Four: Students are initiators and advocates for their own learning.

As students collaborate to build ideas together, they learn to pose their own questions and challenges. Students might ask questions such as the following: Will this pattern work with a different kind of number? What if I try it with very large numbers or very small numbers? Will it work the same for odds and evens? In this way, students become initiators of their own learning.

Students also advocate for themselves by identifying and articulating their confusion. When students engage in discussion with the expectation that they can understand, they learn to pause the discussion if they don't understand. Such pauses are recognized as a contribution to the discourse, allowing all community members to dig deeply and to explain their ideas more clearly. In these ways, students become advocates for their own learning while enhancing the class's collective agency.

Part Four illustrates how each student takes on the responsibility to be an active member of the mathematics community *and* an advocate for their own learning. The chapters include discussion of such themes as strategies a teacher might employ to encourage students' mathematical curiosity while maintaining the coherence of the lesson and what both the teacher and students gain by taking time to address a student's expressed confusion.

The Math

As we examine mathematics community, we give equal weight to "mathematics" and "community." While there are many discussions in the field of education that focus on equitable participation and the development of community, the subject and content of mathematics is sometimes an afterthought at best, or entirely absent, with primary focus put on literacy, history, or science. Is this because classroom mathematics is still seen as the learning of facts and algorithms but not a domain of ideas? Is it because there is still a persistent, perhaps unconscious, belief that some students are wired to do well in mathematics, but others are not?

In the video clips you will be studying, the mathematics students are engaged in centers on the core curriculum content of basic operations—addition, subtraction, multiplication, and division—but it is about formulating and investigating generalizations rather than explicitly learning strategies for solving individual arithmetic problems. The development of accuracy, flexibility, and fluency in solving arithmetic problems is, in itself, an important and fascinating topic. What is also important is for young students to dig even more deeply into the basic operations, through three practices that are highlighted in these videos: noticing patterns and regularities in the number system, conjecturing about what is general in those patterns, and creating a variety of representations to show why and how those patterns hold.

For example, when first graders encounter the operation of addition, they naturally begin to notice and describe patterns. Imagine a group of students generating addition combinations that make 10. A student says: *4 plus 6 makes 10, so 6 plus 4 has to make 10.* Another student chimes in: *I have another* one—*8 plus 2 is 10, so 2 plus 8 works, too.* And a third student says, *but 5 plus 5 is in the middle so it doesn't have a turn-around.* These students, at the beginning of their mathematical journey into addition, are already starting to notice something that is true about addition *in general*, and they have invented a name, "turn-arounds," for their idea (what they will, in later years, encounter as the commutative property of addition). What if the teacher were to challenge them further to state precisely what they mean by "turn-arounds"? What if the teacher were to ask, *Does this work just for these numbers, or does it work for other numbers, too? How do you know?*

In this book, you will see young students investigate patterns in the number system with intelligence and enthusiasm. When given the opportunity by teachers who show genuine interest in their ideas, students enact their intellectual agency by bringing to these problems their own ways of thinking, and they express their identities as they develop and explain their own models, pictures, and diagrams. The teachers in this book assume the brilliance of their students (Aguirre et al., 2024; Delpit, 2012; Leonard & Martin, 2013; Lewis, 2018; National Council of Supervisors of Mathematics & TODOS, 2021) and support them to develop their intellectual power.

The mathematics content in this book, then, is about the core idea of *generalizing*: finding and proving what holds true across multiple, related examples. The generalizations explored by students in the videos are fundamental to a complete and deep understanding of the operations and connect elementary arithmetic to later study of algebra. Just as important, the mathematical practices that are part of generalizing—noticing patterns, articulating conjectures, representing how and why a general pattern holds—create a fertile context for the development of collective mathematical agency. It is content that has many entry points and can be accessed through multiple modes of participation. You can find a summary of the generalizations that students work on in the examples in this book in Appendix A. Teachers in the video were working from lesson sequences which we wrote in collaboration with another group of teachers. Interested readers can find the full lesson sequences in the book *But Why Does It Work?* (Russell et al., 2017).

Finally, we want to make clear that the mathematics content of these lessons is only one aspect of students' mathematical study—a deep dive into the structure of numbers and operations. It is not the only kind of mathematics investigation students should encounter. We advocate a rich mixture in students' mathematics curriculum that includes becoming fluent with a variety of calculation strategies, constructing and analyzing geometric objects, collecting and describing data, studying how people from different backgrounds and cultures have used mathematics, and undertaking projects to answer questions about the world using mathematics (National Council of Supervisors of Mathematics & TODOS, 2016). Important work is being done, including by contributors to this book, to create investigations in which students use mathematics to interrogate issues coming up in their own communities (e.g., Aguirre et al., 2019; Cirillo et al., 2016; see also Appendix B). All of these experiences help to raise students' voices with an emphasis on making sense, being curious, asking questions, and taking risks.

In this book, we are exploring the ways in which teachers support every voice in their classroom to engage with ideas about numbers and operations. These investigations are not limited to calculation but, rather, challenge and inspire students to delve into underlying mathematical structures.

How to Use This Book

Each chapter of this book focuses on an aspect of building a community that weaves deep mathematics with equitable participation and raises questions for reflecting on practice. The following suggestions will help users of this book to get the most from reading and viewing.

Do the Math

Because the mathematics of noticing, conjecturing, and representing general ideas about the operations may be unfamiliar, this resource offers readers an opportunity to investigate some mathematics for themselves as a way to introduce the concepts that students in the video are working on. We strongly suggest engaging with these investigations before watching the related videos in order to get a sense of the complexity of the ideas with which students are working and to better interpret what the students are saying and doing. (If you are like us, you may enjoy adding some math to your day!)

Watch the Classroom Video

One to three two- to eight-minute classroom video clips are the focus of each chapter. We can't say this strongly enough: *Watch the video*. It is possible to understand some of what is going on in the classroom by reading the text and transcript, but seeing the students—noticing their gestures and expressions, hearing the confidence or hesitation in their voices, waiting out the silences—are aspects that don't come through on the page.

When you watch a classroom video, it's tempting to quickly take on a stance of criticism, looking for what the teacher "should have done." Keep in mind that none of us viewing these short clips have the teacher's knowledge about their students or the context of the lesson—what happened before or what the teacher intends to do next. Viewing the video is an exercise in close observation: What do you notice about what students say and do? What does that imply about student learning? What do you notice about what the teacher says and does, and what do students say and do in response?

In most chapters, we recommend that you view each video clip twice, using different lenses. The first time through, focus on what students are learning and the way the teacher gives students access to deep mathematics. What are the ideas students are coming up with? How does the teacher respond to these ideas? In the second viewing, think about students' participation. What opportunities and supports enable different students to dig into the mathematics, express ideas and questions, and interact with other students and with the teacher in building ideas? Is there evidence that students are developing identity and agency as math learners? We provide specific Reflection Questions for these two viewings for each clip.

Transcripts for videos appear at the end of chapters for your reference as you reflect on and discuss with colleagues what you have seen. These are not intended to replace watching the video, since the transcript does not capture gestures, facial expressions, the length of pauses, or other factors that may be important in considering what is happening in the class. Further, while the audio of the video is generally good, and one of us was always making running notes of what students said while the other videotaped, it was not always possible to be certain of students' words. In these cases, we did our best to transcribe accurately, but there may be mistakes, and in a few cases, we have written "unintelligible" to indicate that we could not make out what a student said. There are three aspects of speech that you will hear in the video that we did not capture in the transcriptions: (1) some repetitions of words or phrases made by the teacher or student (e.g., both teachers and students often start a sentence, then restart it); (2) nonword interjections, such as "um"; and (3) variations of pronunciation (e.g., for "going to," people often say "gonna" or variations in between the two). The transcripts, then, are aids for remembering what you heard and saw in the video, but they are not substitutes for watching the video.

Read and Reflect on What Others See in the Video

We believe that we learn best in community—a community in which our tentative thoughts are welcome, in which we try to truly hear and understand others' ideas, in which we challenge each other and ourselves to think about our beliefs and actions, and in which we build stronger commitments than we might be able to sustain on our own. In that spirit, we invited the teachers in the videos and also a small number of educators, representing a variety of backgrounds and roles, to provide brief commentaries on the video clips. The six Collaborating Teachers and seven Critical Friends (introduced in the Preface) all bring to this work both a deep interest in mathematics teaching and learning and a focus on lifting up student voices, especially the voices of students from groups who have been historically marginalized. There is no simple list of strategies that, if implemented, will establish a dynamic mathematics community that includes every student. Rather, teachers' *ongoing, persistent, and determined reflection* on their classroom practice and its effects on student learning is the critical factor.

The purpose of the commentaries in each chapter is to open up different perspectives for viewing each video lesson. As the commentators viewed these classroom videos, they made different observations, raised different questions, and took away different ideas to apply to practice. Thinking about the reflection questions that follow each commentary can help you, alone or with colleagues, choose themes to focus on. Some of the commentaries will undoubtedly connect with issues you have already identified in your practice, while others may provide unexpected questions about aspects of teaching, learning, and participation that have been less visible to you.

In our own years of teaching and of collaborating with teachers over many decades, we, the authors, have consistently found that collaborating to reflect on practice is powerful. Teachers can and do raise questions about their own practice alone, but considering the observations and questions of other educators ensures that one's own questions are not restricted to established routines and beliefs. For that reason, we encourage you to find a partner or form a study group to reflect together on what can be learned from the students, teachers, and other educators who have contributed to this book.

Take Next Steps

Each chapter concludes with suggested "next steps" for you to try in your own practice. We have written these suggestions with the hope that you will explore new ways to ensure equitable participation of your students while maintaining your commitment to deep mathematics.

Notes Organizer

The Notes Organizer is an electronic supplement found at companion.corwin.com/courses/equitable-deep-math. It provides the math activity, Reflection Questions, and Next Steps from each chapter, with space to take your own notes.

Facilitator's Guide

The Facilitator's Guide (also available at companion.corwin.com/courses/equitable-deep-math) is designed for those who lead professional development or a study group based on this book. The guide suggests a chapter-by-chapter plan for organizing study group meetings and offers tips for facilitators.

A Note About Some Terms We Use

There are some terms common in the education community that can refer to different things. To avoid confusion, we specify here what we mean when we use the terms *mathematical practices* and *mathematical representations*.

By *mathematical practices*, we mean the working practices of professional mathematicians. A number of different documents, such as the National Council of Teachers of Mathematics *Principles and Standards* (NCTM, 2000) and the *Common Core Standards* (National Governors Association Center for Best Practices & Council of Chief State School Officers, 2010), have offered lists of practices. Our use of the term includes these but is not limited to them. For example, two key mathematical practices illustrated in this book are *noticing patterns* and *formulating conjectures.*

Mathematical representations are physical, visual, or verbal depictions that embody a mathematical object. Mathematical representations may include pictures, diagrams, number lines, graphs, arrangements of physical objects, mathematical expressions, equations, or the statement of a generalization. Because story contexts can carry so much meaning about mathematical relationships at the elementary level, we include them as representations as well. *Student-created mathematical representations* are drawings, diagrams, models, story contexts, and so on, that come from students' imaginations, as well as more standard forms of representation, such as number lines or arrays, that students have incorporated into their repertoire.

Reflection Question

This resource is structured around four aspects of mathematics community: *every voice matters; collaboration supports student agency; student-created representations offer anchors, openings, and depth;* and *students are initiators and advocates for their own learning.* Think of a recent experience in your own context. How were each of these aspects present? If you were to choose one on which to focus right now in your own work, which would you choose and why?

Part One

Every Voice Matters

First-grade teacher Isabel Schooler describes different ways students have "voice" in her class: They can ask questions, offer what they notice, draw or build clear representations, restate others' ideas, ask for clarification, agree or disagree, suggest other ways, give a partial answer, or provide mathematical language.

All of these modes are ways for students to enter the conversation, to make their ideas known, and to be listened to and appreciated for their thinking about mathematics. "Voice" is not always oral. As Ms. Schooler suggests, students can contribute to discussion by drawing or building a representation to show their thoughts. They might also use signals, like the hand motions you'll see in some of the videos that indicate "I agree" or "I'd like to build on that idea." Even a simple gesture, like pointing to a part of a representation, may allow a student to give voice to an idea.

The chapters in Part One focus on how teachers support each student to develop and use their voice in mathematics.

The following are the major themes of Part One:

- A mathematics community that is focused on deep mathematics and equitable participation allows all students to develop agency as mathematics learners.

- Discussion that is focused on deep mathematics involves a nexus of ideas, offering students different entry points and different modes of participation.

  - To facilitate such discussions, teachers must develop a set of skills and dispositions that promote student engagement.

  - To participate in such discussions, students must also develop a set of skills and dispositions.

In Part One, through viewing videos, reading what our Collaborating Teachers have written, and considering ideas and questions posed by our Critical Friends, we'll delve into the complicated work that teachers do to listen to and sustain every student's voice, while weaving all of those voices into the development of rigorous mathematics for the whole class.

Chapter 1

Creating Multiple Openings Into Engaging Mathematics

Our work is based on two assumptions:

1. Mathematics is an interwoven network of ideas.

2. Students come to school with mathematical ideas, and part of the work of the teacher is to draw out those ideas and help students develop them further.

Students in the elementary grades are learning not only mathematics content but also what mathematics is and what it means to be mathematical thinkers. We view the learning of mathematics as an active endeavor—the construction of ideas, rather than passive absorption of delivered knowledge. As students find their place in the mathematics classroom, we want them to learn that they are capable of making sense of complex mathematical ideas. We want mathematics classrooms to be settings where students have the opportunity to develop positive and productive mathematical identities as doers of mathematics. We hope for mathematics classrooms that are contexts for developing mathematical agency, where students learn to investigate mathematics, share their own ideas, interact with classmates' ideas, and take responsibility for their own learning.

Given this description of what a mathematics classroom can be, we pose, again, the question we asked in the Introduction: How do we ensure that every student contributes a voice to a mathematics community, including students who have been historically marginalized in mathematics, students who have not believed they have mathematical ideas that are important to share, or who, when they have tried to express their ideas, have not been heard?

In this chapter, you will

- do some math for yourself to become familiar with the ideas you'll see students working with,

- watch a video to explore how first-grade students are discovering their own mathematical agency and identities in a whole-class discussion,

- read what our Critical Friends and the classroom teacher have to say about students' learning and participation, and

- consider how to create openings to ensure that every student contributes their voice to the mathematics community.

Whole-class discussion is one of the primary forums in which students are invited to use their voices to contribute mathematical ideas. Take a moment to reflect on your own experiences as a participant in a discussion. Can you remember a time when you have been in a group that seemed closed to your ideas? Did a few people dominate the discussion? Did others seem more certain, more on top of what they had to say? Did you worry that you weren't smart enough or experienced enough to contribute? Or were you usually eager to get out your own ideas? Did you try to understand others' ideas? Was there enough time to hear multiple perspectives?

Now think about what a classroom mathematics discussion might feel like from the perspective of an elementary-school student. Is it a setting in which students are willing to put out tentative ideas, clarify them, ask questions, and build mathematics together? Might it be intimidating for some students? What does it take to create a classroom community that works on deep mathematics and also supports students in participating with all their different personalities, backgrounds, facility with language, and mathematical understanding?

As you consider the classroom video and the commentaries in this chapter, we'd like you to think about three aspects of the principle that *every voice matters*. These three aspects of building community lay a strong foundation for interweaving rigorous mathematics and equitable participation.

1. The mathematics content is powerful, engaging, and challenging, but there are many entry points into the ideas.

2. Accessing the depth of the ideas takes time. "Productive lingering" on a few related questions allows students to dig deeply into mathematics concepts.

3. Teachers can create multiple openings into the mathematics. Students can have voice in different ways.

Do the Math

As we explained in the introduction, you'll get more out of the children's thinking in the video if you do some related math work before viewing. Even though you may find the mathematics itself familiar, doing some work of your own with the same mathematical ideas with which the students are working will help you understand its complexity and importance.

1. Make a drawing or diagram or build a physical model for each of these equations:

$$3 + 6 = 9$$
$$9 - 3 = 6$$
$$9 - 6 = 3$$

Can you draw or build a single representation that shows all three of these equations?

Explain to a colleague how your representation shows all three equations.

2. The three equations are an example of a big idea about the relationship between addition and subtraction. Can you write an "if . . . , then . . . " sentence that expresses the general relationship among these three equations?

If . . . , then . . . .

Can you use your representation (drawing, diagram, or physical model) from Question #1 to explain why your "if . . . , then . . . " sentence is true?

3. Share your "if . . . , then . . ." sentence with a colleague. What is the same about your sentences, and what is different? How do your representations demonstrate why your statements are true?

Watch the Video: "Where Do You See the 3?"

To begin to understand what it looks like to create openings for every student, let's watch and analyze Video 1.1. This lesson is from a sequence of lessons in which Natasha Gordon's first graders are investigating the relationship between addition and subtraction, a complicated idea for young children who are new to thinking about these operations. In order to understand the idea, not only must students conceptualize what each operation does, but they must hold in mind images for *both* operations *at the same time* to understand how they are related. At first, students notice this relationship with specific numbers, but then they gradually consider how it applies to all the numbers with which they are working.

In the previous lesson, Ms. Gordon asked students to create a representation for each of the three equations: $3 + 6 = 9$, $9 - 3 = 6$, and $9 - 6 = 3$. The activity sheet also included the question, "What is the same in your representations, and what is different?"

For the lesson shown in the video, Ms. Gordon selected a piece of student work (Figure 1.1) to project onto the whiteboard and discuss with the class.

Figure 1.1 • A First Grader's Work Representing $3 + 6 = 9$, $9 - 3 = 6$, and $9 - 6 = 3$

1. Draw a representation for $3 + 6 = 9$.

2. Draw a representation for $9 - 3 = 6$

3. Draw a representation for $9 - 6 = 3$

Before the part of the lesson you're going to view, the class talked about what they noticed is the *same* in the three equations and the three representations. The whole-group discussion was paused for a few minutes while students talked with partners about what they noticed—a classroom structure commonly referred to as turn-and-talk—and then students shared their ideas in the whole group. As they spoke about what they saw as similar in the three representations, Ms. Gordon noted their ideas by using colored markers on the whiteboard. She used one color to indicate groups of 3 in each of the representations (shown with gray in Figure 1.2) and a different color to indicate the groups of 6 (shown in black). By the end of the day's discussion, of which this video clip is a part, the whiteboard looked like Figure 1.2.

Figure 1.2 • Ms. Gordon's record of what students noticed about what is the same in the three equations and representations. She used different colored markers for the groups of 3 and groups of 6, shown here as gray and black, respectively.

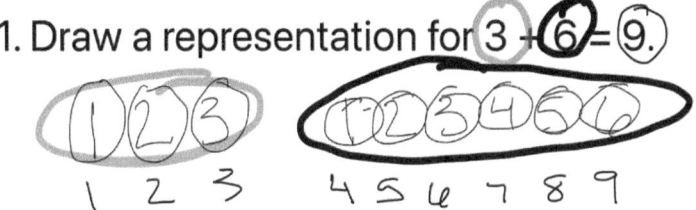

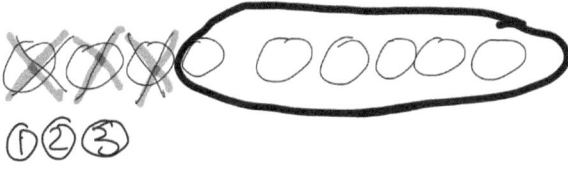

During the discussion about similarities, one student, Livia, pointed out that the numbers are in a different order in the three equations. We'll join the group as Ms. Gordon comes back to Livia to ask her to elaborate her idea.

You'll view the 5-minute video clip twice, with different lenses, in order to help you think about weaving together deep mathematics content and openings for students' voices.

First Viewing of the Video: The Mathematics Students Are Working On

Video 1.1

"Where Do You See the 3?"

qrs.ly/fqfs4v8

To read a QR code, you must have a smartphone or tablet with a camera. We recommend that you download a QR code reader app that is made specifically for your phone or tablet brand.

Watch the video clip, "Where Do You See the 3?," with a focus on the mathematics content students are learning.

Reflecting on the Video: The Mathematics Students Are Working On

[You may want to use the transcript at the end of this chapter as you consider these questions.]

1. What are the important mathematical ideas in this clip?

2. How are students engaging with these ideas? What different ideas are students working on? Are there different entry points into the mathematics?

3. In what ways is the mathematics challenging and engaging for the students?

4. What does the teacher do to focus the discussion and to promote persistence with complex ideas?

Second Viewing of the Video: Finding a Way in Through Multiple Modes of Participation

Watch the video clip, "Where Do You See the 3?," with a focus on how students have voice during this discussion.

Reflecting on the Video: Finding a Way in Through Multiple Modes of Participation

[You may want to use the transcript at the end of this chapter as you consider these questions.]

1. What do you notice about different modes of student participation? Are there opportunities for students to find different openings into the discourse?

2. How does the teacher support students' voices in this discussion?

3. How does the posted student work and Ms. Gordon's use of colored markers support students to find ways in to participate in the whole-group discourse?

4. If you were the teacher in this classroom reflecting on this lesson, what might you want to make note of in order to strengthen student participation? Are there aspects of the lesson that worked well to create openings into the mathematics? Are there questions you have about how you could better encourage students' voices and help students develop agency as mathematicians?

Read and Reflect on
What Others See in the Video

Let's return to the three aspects of the principle that *every voice matters* listed at the beginning of this chapter:

---

1. The mathematics content is powerful, engaging, and challenging, but there are many entry points into the ideas.

2. Accessing the depth of the ideas takes time. "Productive lingering" on a few related questions allows students to dig deeply into mathematics concepts.

3. Teachers can create multiple openings into the mathematics. Students can have voice in different ways.

In this section, you will encounter reactions from some of our Critical Friends to the class session—what they notice and questions they raise. Also included is commentary from Ms. Gordon about how she was thinking about students' entry into the mathematics content during a similar lesson.

1. Critical Friends Consider How These Young Students Encounter the Mathematics Content

"Fact families" are a familiar topic in elementary classrooms. If you search for "fact family" online, you find many worksheets in which children are asked to fill in a series of blank equations with sets of numbers such as 3, 6, and 9. But, without a focus on making sense of these relationships, students might learn to fill in the blanks correctly without thinking about how and why these equations are related.

In Ms. Gordon's lesson, students explore the relationship between addition and subtraction by digging deeply into one example. Young students first need to understand each problem in itself. Many students in first grade are still solidifying the idea that addition can be seen as joining and subtraction as removing, and they are working through what each component of an addition or subtraction equation stands for. Students are also learning what it means to represent a mathematical equation: How do the circles and x's in the drawing represent adding or subtracting? How are the drawings and the equations related to each other? By tracking how each of the three quantities appears in the equations and in the drawings—different ways of seeing how 9 is composed of 3 and 6—students are also taking a step toward understanding more generally how addition and subtraction are related. Because of the many different components of this complex idea, different students might be working on different aspects of the concept within the same lesson.

Here is what some of our Critical Friends have to say about the math in this lesson.

Hetal Patel: The teacher sticks to the math and allows the students to describe it. There's a lot of descriptive language that the kids are using, and she sticks to the words they use. The students have multiple opportunities and a variety of ways to use that descriptive language for what they are trying to make sense of for themselves or for each other.

Virginia Bastable: It feels important to say, Don't wait until you think the idea can be solid for the students. You can help them begin to think about a messy idea. None of the kids seem upset that they're not saying the whole idea. There's nothing negative going on. They're sort of playing with this idea. And even if you're not sure they're going to consolidate it in whatever time period you have, it's still worthwhile. I think that's an important message for people, especially in this day and age when every 20 minutes you have to check off that you met some objective!

Darlene Ratliff: For young students, the mathematics can be murky. Here's how a first-grade class "murks" through it. Nevertheless, the ideas still emerge during the lesson.

 Reflecting on the Mathematics

1. Hetal Patel notices how the teacher makes use of students' own words. Refer to the video or the transcript. How does the teacher encourage and accept students' own language? How does she help students expand and clarify their language, both for themselves and for the understanding of other students?

2. What is your response to our Critical Friends' thoughts about engaging young students with challenging mathematics that is "messy" or "murky" for them or that you may not be able to bring to closure? What do these observations have to do with students developing their identities as doers of mathematics?

2. The Teacher Reflects on a Similar Lesson: What Are Students Learning?

The following year, Ms. Gordon taught the same lessons in her first-grade class. Students represented and discussed the three equations involving 3, 6, and 9 in a lesson similar to the one you watched on video. Based on what she observed in that lesson, she planned the next lesson to focus on two related story problems, using the numbers 6, 9, and 15. This is an example of "productive lingering"—investigating a small set of related problems thoroughly in order to dig into significant mathematics, in this case, the relationship between addition and subtraction. Rather than practicing with pages of sets of problems similar to 3 + 6 = 9, 9 − 6 = 3, and 9 − 3 = 6, students have the opportunity to examine another single example—allowing for time to draw, discuss, and ask questions. The ideas are developed across several lessons. Afterward, Ms. Gordon reflected on the students' representations and discussion of the story problems.

Natasha Gordon: In the session about 3, 6, and 9, students stated that the subtraction equation MUST have a specific answer because of the similarities in numbers in the addition equation. I wondered whether they understood what was happening with the numbers as they related to the actions of addition and subtraction. Going into the next session, I was curious as to whether students would be able to better articulate their understandings and connections, as well as if any new ones would emerge.

For this session, students drew representations for the following problems:

1. *At recess, 6 children are playing on the structure, and 9 children are playing tag. How many children are playing?*

2. *There are 15 children on the playground, 9 children leave. How many children stay on the playground?*

Students were asked whether the first problem helped them solve the second problem and, if so, how?

When we came together to analyze one student's work (Figure 1.3), the discussion was filled with many student voices as we tried to make sense of each problem and each representation. When I asked whether the first problem could help you solve the second problem, I heard the following comments.

Figure 1.3 • A Student's Work Illustrating Two Related Story Problems

1. At recess, 6 children are playing on the structure, and 9 children are playing tag. How many children are playing?

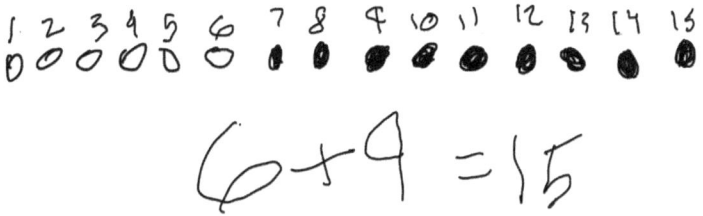

$$6 + 9 = 15$$

2. There are 15 children on the playground and 9 children leave. How many children stay on the playground?

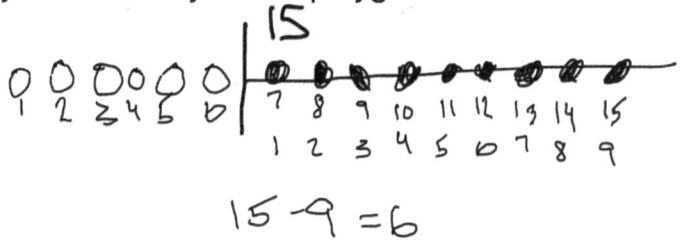

$$15 - 9 = 6$$

Reynald:	Yes, because 6 + 9 = 15 and 15 − 6 = 9, because if you have 15 and you take away 9, it will equal 6 because 6 + 9 = 15.
Ms. Gordon:	Can someone add onto Reynald? How is that helpful if you know 6 + 9 = 15? He said then 15 − 9 will equal 6. How is that helpful?
Xavier:	Both problems, both have 15, but on the bottom one of the problems, the other 9 was taken away, and the first one they were still there, so they went somewhere else.
Ms. Gordon:	Who can say more? Xavier pointed out that there were 15 children in both problems. We can see the 15 circles in both problems, and also I see 15 in both equations. But you said in one problem the 9 is taken away? And then in the other problem what happened with the 9?
Xavier:	The 9 are still playing.
Ms. Gordon:	OK. So in one problem the 9 are taken away, and in the other problem the 9 are still there. Someone else say more . . .

Tatianna: I was talking about this with my talk partner because it *does* make sense because this one is starting with 6 + 9 and it equals 15, and if it's 15 and it takes away 9, it will leave back where we started with 6.

Ms. Gordon: Interesting. Something about this problem [points at top problem and underlines the 6 in the equation 6 + 9 = 15] we *started* with 6. And you said this problem [points at bottom problem and underlines the 6 in the equation 15 − 9 = 6] we're *left* with 6. So how is that helpful? Someone, add on even more. Thank you, Tatianna. This problem we started with 6, and we ended with 15 children altogether. And with this problem, we started with 15 children, and we're left back with 6, Tatianna said. Miriam, do you want to add on?

Miriam: I just want to say it's like you're putting the stories together because if 15 children were playing on the playground, and 6 were playing on the structure and 9 were playing tag, you can put both stories together by, after a little bit of them playing, 9 children left.

Ms. Gordon: But can you say more? How can you put those two together? So you said after some time . . .

Miriam: It's like, it's basically like 9 children come in on the playground and then those 9 leave.

In this five-minute dialog, four students built on each other's ideas and collectively found a connection between the two problems: that the nine students who came onto the playground to play tag could be the same nine students that left the playground, leaving us with the original six children. Tatianna stated that "it does make sense" as she added on to her peers' ideas and tried to explain the connection further. Her exclamation was a sign of her bringing the math together and tying it into the context of the problem. I do not believe all students have arrived at this understanding, but I do believe that as this relationship is explored further, more students will approach this understanding.

 **Reflecting on Ms. Gordon's Writing**

1. As Renald, Xavier, Tatianna, and Miriam discuss the problems about children on the playground, how do their comments build on one another? What are they saying about the relationship between the two problems?

2. What is your response to Ms. Gordon's last sentence in her comments, "I do not believe all students have arrived at this understanding, but I do believe that as this relationship is explored further, more students will approach this understanding"?

3. Critical Friends Consider How Students Are Offered Multiple Modes of Participation

A teacher's responsibility to promote equitable participation is complicated. When using video clips in this book to reflect on our own practice, we are seeing a very small slice of classroom activity. We do not know what the teacher knows about what came before what we see, how the teacher is supporting the needs and strengths of individual students, or how the teacher plans to follow up. Some aspects of what happens on the video are invisible to us. For example, when Ms. Gordon later reflected on the lesson, she commented, "When we broke into turn-and-talk, I made a point of visiting pairs of students who had not yet spoken up. I noticed that four of those students subsequently contributed to the whole-group discussion."

We use these brief video clips not to comment on what the teacher "should have done," but as learning tools for ourselves—to reflect on what we see, to notice which teacher moves bring students into the conversation, to ask questions and wonder what we might do to ensure equitable participation in our own contexts. Even without knowing all that the teacher knows, we can use these short excerpts to provoke questions for our own practice, such as these that our Critical Friends and Ms. Gordon's second-grade colleague, Quayisha Clarke, raise. Note that they viewed the full 30-minute lesson from which this clip is taken.

Quayisha Clarke: Within the turn-and-talk and within the discussion, Natasha [Gordon] would ask questions like, "So what do you think about Livia's idea?" It's really evident to the students that their ideas matter because the teacher is listening to them, and so are the students. What everyone says matters because they're going to talk about this new thing that just happened. I think that gives a lot of power to student voices. I counted, too, and I think it was almost everyone except two students who talked in the whole-group discussion. That doesn't just happen. I'm not going to just walk into any classroom and see all of the students talk in a 30-minute math discussion. She's given access to everyone's voice, and power, as well.

Virginia Bastable: When Ms. Gordon does call on people, she gives the students a lot of time to talk. Even if they're stumbling or not saying their ideas well, there's clearly time for them to get through that, and sometimes they do and fix it and sometimes they say, come back to me, I can't finish. I thought that was supportive of increasing voice.

Darlene Ratliff: As one often sees in first-grade classrooms, boys fidget and move all over the place in this lesson. But they're still engaged. I want to emphasize that they're still engaged, even though they are moving around and fidgety. The teacher didn't spend all of her time saying, "Come on, sit down, stop rocking, stop playing." It was all about the mathematics and the discussion, and not about who's fidgeting. Whether they're fidgeting or not, if they're still tuning in to what you need to have happen, that's an important place for those students.

I also want to say that sometimes voice isn't talking. Sometimes, voice is signaling. As we watched this class, I can pretty much say all of them displayed voice in some way, whether in the turn-and-talk or gestures, everyone had some voice. Some had more voice than others, but overall, there was an atmosphere of voice.

Lynne Godfrey: I was looking at the different kinds of questions the teacher posed. It's one thing to go up to the board to point. But I'm wondering, who gets to answer the thinking questions?

Cindy Ballenger: Josiah was so clear and articulate. It was my feeling that he was helping everybody. So then there's the question, Does the teacher always use Josiah in this way, or are there other kids who take on this role? There might be moments when Josiah is the guy the teacher needs to go to, but you have to take a long view.

Yi Law Chan: I noticed a core of eager hands, the very eager waving to signal something they very much wanted to share. I'm also thinking about how the role of those kinds of signals plays out in a classroom that also has other kinds of signals, as well as opportunities for students to do the turn-and-talks. How will these different nonverbal means of communicating impact different students? What I noticed is that there are definitely students along the periphery, physically and also in their involvement in verbalizing, in the whole group. The turn-and-talk allowed every pair of students to verbalize thinking, but I did notice that there's a central group of students who had the floor more often in the whole-group conversation. So I have curiosity around the voice there in the whole-group setting. How do we assess the impact of a student's contribution on their own learning or the impact of others' comments on their learning? It's a question I have that I'd like teachers to think about. When I visit a classroom, I might not be able to assess this impact, but teachers in the classroom have the ability to assess that over time.

Virginia Bastable: Paying attention over time is such an important theme for a teacher. In any one clip, we're seeing so little that it's hard to make these judgments. But as a teacher in a classroom, I need to notice what kinds of questions I am asking and to whom and make sure those thinking questions get more distributed.

Reflection Questions

1. What do you notice in the clip from Ms. Gordon's class about how students have voice? Are different modes of participation evident? Is it important for every student to speak in whole group? How else might students productively participate in the whole-class discussion? What do you think, in your own context, about Darlene Ratliff's statement that "Sometimes voice isn't talking"?

2. While many students have the opportunity to participate in different ways, in this whole-group discussion, there may still be students who do not find an opening to participate, students who, as Yi Law Chan mentions, are on the "periphery." You may have noticed students in the video—and it is likely this would be true in any classroom—who don't seem to actively participate. What would you want to know about these students? How might you follow up with them? What might you plan in order to provide openings for them to engage in the next whole-group discussion?

What Do You Want to Remember From This Chapter?

Take a few minutes to note for yourself ideas you want to hold onto as you continue to investigate the meaning of a mathematics community and how to build it. What teacher moves have you noticed in this chapter that you want to bring into your own practice? Here are some of the ways we like to think about what the teachers and Critical Friends in this chapter have said about creating openings for every student:

- **Create multiple entry points and openings.** A large part of equitable mathematics teaching is access. Give students openings to enter into mathematical ideas through a variety of representations, including words, equations, and diagrams, to ensure the greatest possible participation.

- **Create an expectation of "productive lingering" on important ideas.** Structure lessons to focus on one or a few related questions. Encourage students to investigate those questions deeply, welcoming questions and ideas from many voices.

- **Celebrate different forms of participation in class discussion.** Participation in math discussion can include stating what you notice, asking questions, gesturing, building on classmates' ideas, and indicating agreement, disagreement, or confusion. Make openings for, acknowledge, and celebrate all of these forms.

- **Pay attention over time.** Develop ways to track which students are participating and in what ways. Who speaks up in whole-group discussions? What kinds of questions are you asking to which students? Who is sharing their work? Who is commenting on other students' ideas or representations?

Taking a Next Step

List the students in your class and place their names in one of three groups: (1) one-third of the students who participate most in math class, (2) one-third of the students who participate least in math class, and (3) those who fall between the two. As you look over the names in the three groups, what do you notice? What questions do your lists raise for you? If you don't have your own classroom, adapt this activity to another group with which you are working.

Video 1.1 Transcript: **"Where Do You See the 3?"**

Ms. Gordon: Livia, I'm going to come back to you because you started talking a little bit about how they're different. Can you say more about how they're different? What were you saying?

Livia: So what I was saying was that they're the same numbers but they're switched up, like in the first equation it says 3 plus 6 equals 9. And then the next equation is 9 minus 3 equals 6. And then the next is 9 minus 6 equals 3.

Ms. Gordon: How many friends also see that, that they're the same numbers but they're a little bit different because they're in different orders? For example, Caitlyn started talking about the number 3. Where do you see, in the first representation, where do you see 3? Audrey?

Audrey: The first 3 in the first one.

Ms. Gordon: The first three circles. But in the second representation, where do you see the 3? J'aimeson?

J'aimeson: [Points to board.]

Ms. Gordon: So, these three right here. Audrey, can you say a little bit more? What did you just say?

Audrey: I said the ones that were taken away.

Ms. Gordon: So how is that different? Just in those first two. For the first problem we noticed that the 3 is represented by the first three circles. But in the second problem the 3 is represented by the three that are being taken away. Josiah?

Josiah: Because that one is adding 3, and that one is taking away 3.

Ms. Gordon: How can you tell that this one is taking away 3?

Tierra: Because it says it's taking away 3.

Ms. Gordon: Where?

Tierra: Right here.

Ms. Gordon: Ah, in the equation. 9 take away, 9 minus 3. How is that represented in the work? What did this scholar do to represent 9 take away 3 or 9 minus 3?

Anu: They x'd it.

Ms. Gordon:	Ah, they crossed them out. Okay, so we talked about 3, we noticed 3 is represented by the first three circles in the first equation; 3 in the second equation is represented by the three that are taken away. What about the third representation? We said we saw 3 there, 3 is in this equation and in the representation, but where is the 3 in this representation? Tunmiche? What does the 3 represent in the third equation?
Tunmiche:	Right there.
Ms. Gordon:	Where?
Tunmiche:	[points to board]
Ms. Gordon:	Can someone say more? Okay, so we see the three are right here. Can someone say more? Josiah, I see that you want to add on.
Josiah:	The 3 is represating for what it equals.
Ms. Gordon:	OK, what it equals or . . . ?
Anu:	The answer.
Ms. Gordon:	Who can say more—what it equals, the answer . . . Audrey?
Audrey:	Like what it means.
Ms. Gordon:	Josiah?
Josiah:	The total.
Ms. Gordon:	Okay? Interesting, you're saying the total. When we say total, the total is how much we have . . .
Students:	Altogether.
Ms. Gordon:	Is the 3 what we have altogether?
Students:	No
Ms. Gordon:	So what is this 3? Janiya?
Janiya:	What we take away?
Ms. Gordon:	Okay, so we know this problem is subtraction. We are taking away, but what does this 3 represent?
Anu:	What remains?
Ms. Gordon:	So, in the first problem the 3 is what we started with. In the second problem we took away 3. And in the last problem 3 is what we're left with.

Chapter 2

Establishing Expectations and Structures for a Participatory Mathematics Community

In a third-grade classroom, about three weeks into the school year, the teacher asks students to write down one thing that they like about math discussions and one thing that they don't like or that makes it hard for them to participate. The teacher then asks if any students would be willing to share what they wrote:

Clyde: Sometimes, my tongue gets tied up. I know what I want to say, but it isn't coming out right. [A few other students nod in recognition.]

Teacher: Is there anything that helps or doesn't help when that happens?

Clyde: When people start waving their hands, I feel like I'm not talking fast enough.

Lizzie: Sometimes I feel like that, too. It seems like everyone else has the answer, and I don't.

Sarita: It helps me if people say they agree or they want to add on to what I said. Then I can tell they were listening to me.

Enacting the expectation that every student's voice matters in the mathematics community requires efforts that go far beyond good intentions. In classrooms in which every student is given time and space to contribute ideas, messages to students about inclusion are explicit, and students who may be reluctant to contribute are supported with determination and consistency. Students need to know, from the very beginning of the year, that the entire community is responsible for making space and time for every student to be part of the building of ideas.

Chapter 1 focused on different ways of creating openings for students to find their way into engaging mathematics—through words, gestures, creating representations, and explaining other students' representations. What expectations and structures can teachers establish in their classrooms, right from the beginning of the year, to ensure that every student—those from groups that have been historically marginalized in mathematics, those who need time to get their ideas out, those who, for whatever reason, are hesitant to speak at all—is heard? How can a foundation be laid to ensure that, when students do contribute, both the teacher and their peers are listening?

In this chapter, you will

- consider how teachers establish, from the beginning of the year, expectations, commitments, and classroom structures that support respect for the contributions of every student, and
- watch two video clips, one of a first grade and one of a third grade, enacting one classroom structure designed to support all students to express their thinking.

As you watch the classroom video and read the commentaries in this chapter, consider these three aspects of the principle that *every voice matters*:

1. Developing curiosity about students' thinking is fundamental to providing space and time for every student.

2. Getting started at the beginning of the year includes establishing class expectations and commitments. In order to enact and maintain those commitments, the class periodically reflects on them.

3. Students require time to practice using classroom tools and structures that support equitable participation.

Getting Started at the Beginning of the Year

In the video of Ms. Gordon's first grade from Chapter 1, students were well on their way to knowing how to offer ideas, listen to their classmates, and comment on each other's thoughts. Since the beginning of the school year, Ms. Gordon had been communicating to her young students that she values their thinking, including partial understandings, and that they are expected to listen to and build on each other's ideas.

Each school year, in their very first mathematics class, our Collaborating Teachers begin to establish expectations for equitable participation. They emphasize that this work continues throughout the school year as students learn what these expectations look like in action and what their responsibilities are toward their own and others' participation. In discussions with our Collaborating Teachers and Critical Friends, three critical aspects of what teachers do to establish mathematics community stood out: developing curiosity about students' ideas, establishing norms, and introducing and practicing structures that support participation.

Developing Genuine Curiosity About Your Students' Ideas

Trusting that every student has, and can develop, ideas about mathematics is crucial to equitable mathematics classrooms. And yet we have all had students who at first baffled us, whose thinking we found hard to grasp, who had us wondering whether they were making sense of mathematics in any way. Have you ever heard a colleague say—or maybe even found yourself thinking—this student "doesn't have a math brain," or "this student can only memorize procedures," or, simply, "this student is so confused"? All of these, and many other similar generalizations, act as barriers that distract us from drawing out and building on a student's ideas. Furthermore, students may hide their ideas. They may be reluctant to express something they think might be wrong, or they may have a hard time finding the words to say what they mean in a way that allows others to follow their thinking. They may have different ideas woven together that require some teasing apart. Often, students have beginning, partial understandings as they are making sense of new ideas. But having emerging, incomplete understandings is, after all, what it is to be a learner. Finding the sense in what each student is saying involves listening to, through, and underneath what a student expresses in their work, sometimes restating to check whether you've got their idea right, or asking questions based on what the student is expressing rather than on the answer you are hoping for.

It's easy to give in to calling on the students who are reliably articulate and whom you've come to expect will push the math ideas forward. But if you learn to nurture productive contributions from only certain students, you are inevitably training yourself not to expect useful contributions from others. This is an insidious, often unconscious, cycle that happens at times to all of us. It is a profound part of the work of teaching to maintain one's curiosity about all students' ideas, to learn to listen for what is relevant and important, and to help students learn to listen in the same way.

In her classic essay "On Listening to What the Children Say," Paley (1986) chronicles how hard she had to work to develop her own curiosity about children's ideas by holding back on interjecting her own: "When my intention was limited to announcing my own point of view, communication came to a halt" (p. 125). She asked herself, "When did my words lead the children to think and say more? . . . The decisive factor for me was curiosity" (pp. 124–125). Becoming curious doesn't simply happen with a decision to do so. Teachers need to practice developing their own curiosity and sense of wonder about the ideas of every student. When this genuine curiosity is communicated to students, they learn that their ideas matter and that the effort they make to express them is received with interest and respect.

Here is what some of our Collaborating Teachers and Critical Friends have to say about developing curiosity.

Virginia Bastable: To develop curiosity takes time. It isn't that you either do or don't believe that kids have ideas on their own. A teacher needs to feel safe enough to try some things with students and begin to notice what they're capable of. You might not start with genuine curiosity, but it develops as you provide openings to see what students can do.

Isabel Schooler: It makes kids feel special and that their work is important when you say, "Wow, let's think about that."

Cindy Ballenger: It's interesting to listen to Natasha Gordon [in Chapter 1] and wonder about when she's genuinely curious, because her manner and intonation beautifully communicate curiosity—it's wonderful. But probably she sometimes really is surprised and wondering and, I think, sometimes probably not. It complicates the idea for me to think about that and what it means to develop and communicate curiosity.

Establishing Norms or Commitments Around
What It Means to Contribute and What It Means to Listen

Establishing agreed-on guidelines for participation in mathematics discussions, making these explicit and public, and keeping them fresh throughout the year is another part of the work in welcoming every student's contributions. After all, students may never have participated in a math discussion before, may be unaccustomed to articulating confusions or partial understandings, and may not know how to listen to classmates.

To get started establishing a set of classroom commitments, ask students to reflect on what helps them to participate in math class. Pose specific questions: What helps you feel like you can say something in math? What do you do when someone else is sharing their ideas? What would you do if you don't agree with something someone says in math? How would it be helpful to you to hear that someone doesn't agree with something you have said? It may be best to have several discussions before actually trying to write a class list of guidelines so that students have had some time to think about what is important to them in the context of actual experiences. Teachers suggest that the class try to capture what is most important to them, and that a list of commitments be relatively short so that the major points can be kept in mind without being overwhelming.

In order for such a list to be useful, students need to reflect on how their commitments are working in practice. An example of an agreement can be pointed out in real time by the teacher. For example, if one of the class commitments is, "When I don't understand what someone is saying, I can ask a question or say what I think the idea is," the teacher might point out an instance of this right after it occurs: "I noticed that when Robbie didn't understand what Sara was saying about adding 1, he asked her a question. Robbie was listening carefully and trying to understand her idea." It's important for students to periodically reflect on how their class commitments are working and whether they need to add to or modify them. This kind of reflection best happens immediately after a mathematics class, so that students still have in mind how they shared their own ideas and how they received others' ideas.

During her early years as a teacher, working closely with her coach, Liz Sweeney, Isabel Schooler discussed classroom norms with her first graders and created the poster in Figure 2.1.

Figure 2.1 • Example of Class Agreements

Community of Mathematicians

▶ Everyone's voice matters
- Ask questions   · Give partial answer
- I notice...   · Mathematical language
- Draw or build a clear representation
- Restate others' ideas   · Agree/disagree
- Ask for clarification   · Suggest other ways

▶ Everyone engages with the work

▶ Everyone collaborates

▶ Everyone is self-reflective

In the next chapter, we'll view a video clip from Ms. Schooler's classroom, illustrating how "everyone's voice matters" was enacted in the case of a particular student.

Here's what some of our Critical Friends have to say about establishing classroom agreements:

Marta Garcia: In my work with teachers, we try to establish the difference between *rules,* like when to sharpen pencils or ways to move in the classroom, and *agreements* or *commitments* about how we will support each other's learning and have our own learning supported. It's not enough to simply post "Classroom

Agreements" on a laminated poster, which seems to suggest that the agreements are imposed. Rather, the class coming up with their own ideas is what matters. Sometimes teachers tell me that at the beginning of the year, students don't have any ideas about what it means to learn together. But after the class has engaged in math activities and discussions, those opportunities to talk and develop commitments can evolve over time. Teachers can include opportunities for students to share how they are experiencing math class, paying attention to how status is playing out in their classrooms. For example, how are the classroom commitments honoring the varied contributions of students' language and cultural competencies?

Hetal Patel: I think coming back to the norms regularly is key, both having students reflect on them and the teacher pointing out what students did or said that connected back to the norms. I've also talked with teachers about ways to think about commitments as responsibilities. There are many layers of responsibilities, for example, responsibilities for individual students for their own learning, partners' responsibilities for one another's learning, and responsibilities to the whole class as the larger community of math learners.

Yi Law Chan: Here's some of what I've noticed in classrooms where teachers create a safe environment and where students are honored and respected. Expectations for ways of engagement are explicitly named and positively reinforced when the named expectations occur in the classroom. Teachers are relentless and consistent when it comes to things such as wait time or checking for understanding (raise a quiet thumb if . . . ; take a minute to think to yourself, then turn to your partner, etc). Invitations for students to self-reflect on their engagement with norms/agreements/commitments raise awareness of these expectations. This is a sort of mindfulness exercise for how we operate in math class. Teachers ask, "How did doing [insert something from the norms/commitments/agreement chart] help you learn today? Was there a norm/commitment/agreement that was hard for you today? How so? How might your classmates (or you) help you work around this?" I think that especially in places where these kinds of behaviors and mindsets are not the norm, it takes extra effort to reset. Teachers are an important part of the feedback loop. They have to not only model and demonstrate their beliefs about how children best learn mathematics but also provide feedback and affirmation to students.

Cindy Ballenger: As a teacher, I relate especially to what Yi Law says about ways of engagement that are both explicitly named and then mentioned when they occur in the classroom. I don't think I personally have ever posted a list of commitments, although I know many teachers who do, but I think it's important to point out and name effective and democratic ways of talking when they occur, for example, listening hard, waiting, disagreeing respectfully. Then these named ways of participating can be referred to and praised in the future as they occur.

Lynne Godfrey: Even if kids have been in classrooms where they have engaged in math discussions on a regular basis before they come to you, it's still a new community at the beginning of the year. You're establishing new agreements and ways of being with each other.

Providing Multiple Tools and Structures for Students' Contributions, and Then Practicing These so That Students Feel Comfortable Using Them

Across the six classrooms you are viewing in this book, teachers established a variety of tools and structures that facilitate openings for student participation. Some of the teachers taught hand gestures that indicate responses to a classmate's idea, including "I agree," "I don't agree," and "I want to build on that idea." Some teachers encouraged students to use sentence starters, similar to the hand signals, so that they would start their own contributions with "I agree with Deanna because" or "I want to add on to what Alex said."[1] All of the teachers used the turn-and-talk structure, in which, during a whole-group discussion, students are asked to talk to a partner for a short period of time about a question or idea.

All of these techniques, used well, are designed to make openings for more students and to encourage students to listen to and build on each other's ideas. In recent years, "sharing strategies" has become more prevalent in the mathematics classroom, replacing the almost silent math class some of us

---

[1]Some of the teachers had read together the book *Academic Conversations* (Zwiers & Crawford, 2011), and were applying its principles and techniques in their classrooms. The book does not include anything about the mathematics class specifically, but the teachers thought through how to apply its principles to math class.

experienced in which students speak only to provide answers to a problem. However, "sharing strategies" can take the form of students speaking one after another, each eager to report their own approach to solving a problem, but with no connections made between their contributions. These serial presentations don't encourage students to listen to, try to understand, and connect with each other's ideas. The hand gestures and sentence starters emphasize this connection. Starting one's contribution with "I agree with" or "I want to build on" involves practicing how to listen to a previous speaker's idea.

What enables such techniques to be adopted successfully so that they support productive interactions? Any technique can be used in a rote manner, might be adopted only by the students who would be eager to share their ideas anyway, or might just fall away and not become a routine part of classroom discourse. Turn-and-talk, for example, is a structure that puts the responsibility on every student to speak with their partners. But sometimes students simply remain silent, or only one student does all the talking, or students talk about topics other than the math idea they're supposed to be considering. How do teachers establish turn-and-talk so that these conversations work to involve students more fully in the mathematics and help them bring new ideas back to the whole class?

Watch the Videos: Turn-and-Talk, Grade 1 and Turn-and-Talk, Grade 3

Let's pause here to watch and analyze two brief clips of turn-and-talk as examples of this participatory structure. The first is from Natasha Gordon's first grade. You met Ms. Gordon and her students in Chapter 1 working on the relationship between addition and subtraction. This one and a half-minute clip occurs later in the year when students are working on another set of ideas. Ms. Gordon has posted a student's representation of two related story problems and poses the questions, "What does this work tell us? What do you see?" Then she asks students to talk with a partner.

The second two-minute clip is from Emmanuel Fairley-Pittman's third grade class. (You will see more of his classroom in Chapters 6 and 7.) The class is comparing $8 - 3$ and $8 - 4$ in order to delve into the following question: given a subtraction problem, what happens to the difference when the subtrahend (the second number in the expression) increases? They have built $8 - 3$ and $8 - 4$ with cubes, and Mr. Fairley-Pittman says, "I want you to talk with your neighbor. What happens to the difference when the subtrahend increases?"

In these two clips, you won't be able to hear most of the words of individual students. Rather, look at students' gestures and expressions as you watch the first graders in Video 2.1 and the third graders in Video 2.2.

Video 2.1

Turn-and-Talk, Grade 1

qrs.ly/ybfs4vf

Video 2.2

Turn-and-Talk, Grade 3

qrs.ly/aqfs4vh

Reflecting on the Videos: Turn-and-Talk

1. What do you notice about how the teachers set up the turn-and-talks?
2. What do you notice about students' gestures and expressions? What do these indicate to you?
3. What are aspects of this routine that students know how to do?
4. What do the teachers do during the turn-and-talks?
5. After the turn-and-talk, how do the teachers' questions connect to what students have discussed?

These clips are just brief examples of turn-and-talk. Across many turn-and-talks in the collaborating classrooms, we noticed a few consistent characteristics that support productive conversations. A good turn-and-talk

- **Focuses on clear, open questions:** The teacher gives students a clear question to consider, one that can be answered with a range of responses. Teachers often ask, "What do you notice?" about, say, a chart or a sample of student work. Or they ask students to consider an idea posed by a student and recorded by the teacher: "What do you think Kenny means? Do you agree and why? Do you have any questions for Kenny?"

- **Refers to something visual:** The question for turn-and-talks often refers to something that has been posted so that all students can see it. This provides a reminder of the mathematics students are considering and gives students something to refer to as they talk.

- **Requires careful listening by the teacher:** The teacher listens to the pairs or small groups, choosing different students to talk with across a lesson or lessons. They especially spend time with students who haven't spoken in the whole group to hear their ideas and think about ways to bring them into the public discussion. This is an opportunity for teachers both to assess student understanding and to demonstrate interest in students' ideas.
- **Stays brief and focused:** Turn-and-talks are often timed so that the discussions stay focused. Students know they have a short time to express their ideas and don't have so much time that they run out of things to say.
- **Has explicit expectations for participation:** The teacher often establishes partners ahead of time as well as clear practices that are part of the turn-and-talk. For example, students are expected to sit up and face each other and to make sure that everyone in the pair or small group has a chance to speak.

These aspects of turn-and-talk are deliberately practiced, and the class reflects periodically on how the turn-and-talks are working and what would improve them.

As the year goes on and the purpose and characteristics of turn-and-talk become routine, some of these conventions may become more flexible. For example, students might choose their own partners. Our Critical Friend Cindy Ballenger offers this bit of advice.

 **Cindy Ballenger:** When you can, observe, in particular, turn-and-talks with students who are multilingual learners or who need more time to get their ideas out. When might it be helpful to structure turn-and-talks so students have more time? What do you gain from shorter turn-and-talks? from longer ones?

The clear assignment, the monitoring and interest of the teacher, and a predetermined number of minutes combine to keep students focused on the mathematics. Over time, as teachers show interest in the ideas they are hearing and sometimes ask students to bring those ideas back into the whole-class conversation, students learn that what they have to say is important.

Whichever structures and techniques you choose, the key to making any participation technique work is continued practice, reinforcement, and reflection. Students need to first learn the technique, then be reminded how to use the technique in context so that it becomes routine, get feedback about when they are using it successfully, and participate in reflection, throughout the school year, about how the technique is working for themselves and for the class as a whole.

Reflection Questions

1. Think about establishing commitments or agreements in mathematics in your own context. Are there class commitments already in use? If so, how were they established? Did students have input into them? Do students know what they are? Do both teachers and students refer to them in context and reflect on how they are working? If not, how might you want to start establishing class commitments? To lay the groundwork for developing commitments, are there discussions you might have with students about what helps them express their ideas in math class and what helps them listen to other students' ideas? What would it mean to be, as Yi Law Chan puts it, "relentless and consistent" in developing, referring to, and reflecting on a set of classroom commitments about participation in math class?

2. What are classroom techniques (hand gestures, sentence starters, turn-and-talk, etc.) used in your own context to encourage student participation in whole-class discussion? How are they working? Are there aspects of these techniques you'd like to reflect on with students?

What Do You Want to Remember From This Chapter?

Let's return to the main ideas of this chapter:

1. Developing curiosity about students' thinking is fundamental to providing space and time for every student.

2. Getting started at the beginning of the year includes establishing class expectations and commitments. In order to enact and maintain those commitments, the class periodically reflects on them.

3. Students require time to practice using classroom tools and structures that support equitable participation.

Think about how these ideas might inform your own practice and take a few minutes to note for yourself ideas you want to hold onto as you continue to investigate the meaning of a mathematics community and how to build it. What teacher moves have you noticed in this chapter that you want to bring into your own practice? Here are some of the moves we, our Collaborating Teachers, and our Critical Friends have identified in this chapter:

- **Build agreements about speaking and listening.** Committing to respect for and attention to every student's math thinking requires explicit, ongoing attention to building class agreements about speaking and listening.

- **Develop participatory structures and practice, practice, practice them.** Find structures like turn-and-talk, hand signals, phrases students can use to respond to prior statements, and others that are a good fit for you and your classroom. Decide which of these to implement. Support students to practice and reflect on how these structures are working for them.

- **Stay curious.** The teacher's expressions of curiosity are key in communicating to students that they have ideas worth investigating and sharing.

Taking a Next Step

1. Make some time to focus on your own developing curiosity about students' ideas. Choose a student whose mathematical ideas you don't know well, don't feel you understand, or are curious about. (You might want to refer to the list you made at the end of Chapter 1.) Over the next week, during math class, check in frequently with the student you chose. Spend some individual time with that student doing a math task you think they will find accessible and enjoyable. Ask open-ended questions (e.g., what do you notice? can you draw a picture that shows what you're thinking?). Listen. Let the student know through your questioning that you are curious about their ideas. Try to articulate to yourself what the student's strengths are in mathematics. Take notes immediately after each of your interactions with the student in order to record as much as you can about what they said and did. You may want to audio- or video-tape some of your interactions, or ask a colleague to take some notes. In choosing ways to record, you'll need to be sensitive to whether they feel intrusive to the student. However, students are often pleased that you are interested enough in their ideas to write them down.

2. Choose one technique or structure and, the next time students are using it, reflect on whether or how it is promoting participation and discussion. Are only a few students using it successfully? What do students need to practice in order to use it more successfully? Have you reflected on this technique with your students lately? If not, you may want to ask them whether and how it is helping them to participate. You may want to record or ask a colleague to take some notes for you.

Chapter 3

Making Space and Time for Every Student

Many students are, at first, reluctant to contribute their ideas in whole-class discussion. There can be many different reasons for this reluctance. Some students may have had classroom experiences in which only a few students, who typically provided the "right" answer, were called on. They may assume their ideas are not of interest. Some students know they have mathematical ideas but are not sure how to get their words out in order to express those ideas in a whole-group setting. Having an idea and articulating it in a way that others can understand it—as we have all experienced ourselves—are two different things. While there are those who seem to be able to fluently articulate what they want to say on their first try, most of us are figuring out the words we want *as we say them*. We typically start and stop, backtrack, repeat ourselves, reword until we find the way to express our thoughts. This may be even more true for young students, who are learning both everyday language and mathematical language, or for multilingual students who are learning the language predominantly spoken in the classroom.

Do we give our students enough time and encouragement to get their ideas out? Do they know that this is an important effort and will be not just tolerated, but appreciated?

In this chapter, you will

- watch a video in which a first grader needs time to express his idea in a whole-group discussion,

- do some math about the relationship between multiplication and division to prepare for watching a second video,

- watch a video in which a third-grade teacher interacts with students who appear to be unsure of the value of their contributions, and

- read a teacher's profiles of three students in order to examine how the teacher integrates mathematical learning and participation goals for individual students with the overall math agenda for the class as a whole.

As you consider the classroom video and the commentaries in this chapter, we'd like you to think about these two aspects of the principle that *every voice matters*:

---

1. Students are learning that everyone has mathematical ideas that are worthy of being heard and considered. Both students who are eager to articulate their ideas and students who need more time or support to articulate their ideas are included and affirmed in class discussion.

2. Within the context of a lesson in which the whole class engages, teachers think hard about how to integrate goals for each student with the math agenda for the class.

Watch the Video: "I Mean Adding"

To open this chapter, we want to watch and analyze Video 3.1, "I Mean Adding," as an example of what it looks like when a student is provided space and time to speak. Isabel Schooler's first-grade class has been working on their class commitments since the beginning of the year (see their poster in Figure 2.1). This clip takes place about six months into the school year.

Like the work in Ms. Gordon's class in Chapter 1, the math focus of this lesson is on the relationship between addition and subtraction. Ms. Schooler starts the class by reminding students about two problems they had discussed previously:

1. Jan is holding 7 red balloons, and Tim gives her 5 blue balloons. How many balloons is she holding now?

2. Jan is holding 12 balloons. 5 balloons fly away. How many balloons is she holding now?

Ms. Schooler draws students' attention to a poster of the two problems that includes representations of the problems students had developed in that prior session. (See Figure 3.1. In the original poster, the balloons were red and blue as described in the problems.)

Figure 3.1 • Two Related Story Problems

Jan is holding 7 red balloons, and Tim gives her 5 blue balloons. How many balloons is she holding now?

○ ○ ○ ○ ○ ○ ○   ○ ○ ○ ○ ○
1 2 3 4 5 6 7   8 9 10 11 12

7 + 5 = 12

Jan is holding 12 balloons. 5 balloons fly away. How many balloons is she holding now?

○ ○ ○ ○ ○ ○ ○   ∅ ∅ ∅ ∅ ∅
1 2 3 4 5 6 7

12 - 5 = 7

She begins the discussion by asking if students notice any similarities or differences between the first problem and the second problem and whether the first problem could help them solve the second problem. When she poses these questions, she notices a student who has never before raised his hand to speak in discussions about noticing patterns in the operations. The 2½-minute video clip shows Ms. Schooler calling on the student, Jeuri (pronounced "Jow-ree"—the rolled "r" sounds almost like a "d"), and what he says.

Watch Video 3.1, "I Mean Adding," with a focus on the ideas that Jeuri expresses, how he expresses them, and how the teacher and the class receive his ideas.

Video 3.1

"I Mean Adding"

qrs.ly/3ufs4vi

Reflecting on the Video: "I Mean Adding"

[You may want to use the transcript at the end of this chapter as you consider these questions.]

1. What is important mathematically about what Jeuri contributes to the discussion? What is he noticing?

2. How does the teacher and the class show support for Jeuri's ideas?

3. Reflect on your own response to Jeuri's contribution. As you watched Jeuri take time to articulate his idea, were you uncomfortable with his pauses? Do you think the teacher should have provided more scaffolding for him? Why, or why not?

Read and Reflect on What Others See in the Video

What makes it possible for all students to have the time and space to participate as full members of the mathematics community? Let's return to the aspects of the principle that *every voice matters* listed at the beginning of this chapter:

> 1. Students are learning that everyone has mathematical ideas that are worthy of being heard and considered. Both students who are eager to articulate their ideas and students who need more time or support to articulate their ideas are included and affirmed in class discussion.
>
> 2. Within the context of a lesson in which the whole class engages, teachers think hard about how to integrate goals for each student with the math agenda for the class.

Our Critical Friends have some comments and questions about how this video clip can inform our thinking. Ms. Schooler also comments on her thinking about this class session.

Critical Friends and Collaborating Teachers Consider "I Mean Adding"

Often students in elementary grades need more time to articulate their thoughts than might be comfortable for their teachers. Sometimes students hesitate, stop in the middle of their thought, backtrack, repeat themselves, break off what they are saying and don't seem to be able to continue, or simply take what seems like a long time to get started, as Jeuri did. As teachers, it can be difficult to balance support for the participation of all students, not just the most articulate, with the natural restlessness and straying attention of a class full of young students. There is, of course, no set of rules to make finding this balance easier. The point is, rather, that both of these goals must be present and in tension—supporting each student to express ideas and keeping the discussion going in a way that engages the whole class.

A key principle here is that finding this balance, and keeping both goals present, is not just the teacher's job. It has to be the collaborative work of the whole community, and for that to happen, community commitments about including and listening to all voices must be established collaboratively and reinforced frequently, as was the case in Ms. Schooler's class.

While we might at first feel that waiting through Jeuri's slow articulation is difficult, we need to help ourselves and our students not to focus on the speed of his speech but to remain eager and curious about what Jeuri has to say. Although we might feel the urge to help Jeuri along, to help him get his words out, his complete contribution actually took only about two minutes— surely not too much time to allocate to this important moment in the group's work.

Quayisha Clarke: Isabel [Schooler] had asked how these two problems are related. She stuck with Jeuri, allowing him to explain what he was thinking. She rephrased the question at one point and then gave him the time and the space he needed to get his ideas across. I had Jeuri the next year in second grade, and I don't think I would look at this clip the same way if I didn't know him as a mathematician. This clip portrays Jeuri's difficulty in communication and how this was such a really important step for him to talk in front of his class, but it might skew what people see as his mathematical ability. Jeuri is very strong in math. What was so powerful was how Isabel and the class could hold him in that safe space.

Marta Garcia: To me it was important that Jeuri had the confidence to want to share and that Ms. Schooler and his classmates gave him the opportunity to get his ideas together. Ms. Schooler says, "You got this," and you can just see the confidence as he turns around and says, "It's addition." You can see that Jeuri had ideas that needed to come out.

Working with teachers, sometimes they say to me, "But it might embarrass the student if they're up there very long" or "I don't want to call on certain students because they're just going to stand there, and then what do I do?" For example, students who are multilingual might be considered reticent to share their ideas when they don't yet have the language to express the mathematical ideas they are considering. Teachers like Ms. Schooler who offer increased wait time and who encourage the use of representations and nonverbal communication such as gesturing support the participation of multilingual students.

Michelle Sirois: One thing that stuck out to me was that after he shared 12 – 5, Isabel [Schooler] continued to push and ask follow-up questions. For students who take longer to share, you might let them share one thing and then think, "Good, they did it, now I'm going to call on someone else to move the lesson forward." I thought it was so powerful that Isabel continued to engage him to fully answer what the question was.

Emmanuel Fairley-Pittman: I think that part of becoming an equitable educator is that you're okay positioning yourself to be uncomfortable. Watching this clip, I'm wondering what people need in order to fully understand the power of giving Jeuri that time and that space, and not just Jeuri, but giving *the class* that time and that space.

The Teacher Reflects on "I Mean Adding"

Ms. Schooler: In the video clip, I called on a student who does not frequently volunteer to speak in front of the class. Whenever I listened in on his conversation during turn-and-talks, he shared his ideas, but it took him a long time to get them out. I knew from his representations that he was seeing connections and could move our thinking forward. Sometimes it would take him some time to form a sentence, but I knew that this opportunity to share his thinking and see himself as a mathematician would be powerful both for his development and for our class. It took this student almost a full minute to get his first thought out. Despite the time it took, our class was quiet, and we gave him that time. One student even said, "We believe in you." This felt like a critical moment for Jeuri and our class. From then on, Jeuri knew that he could go in front of the class with a partial thought or even a full thought that took some time to come out and he would be respected. Our class community learned over time that everyone moves at a different pace and that it is important for all of our learning to give each other time to think.

Reflection Questions

1. Emmanuel Fairley-Pittman comments, "I'm wondering what people need in order to fully understand the power of giving Jeuri that time and that space, and not just Jeuri, but giving *the class* that time and that space." How do you interpret what he means? How might his comment inform your own practice?

2. Quayisha Clarke notes that if she didn't know Jeuri herself, she might view this clip differently: "It might skew what people see as his mathematical ability." What are aspects of how we interpret students' social interactions and modes of communication that might block us from recognizing their mathematical thinking? Can you think of an example of this from your own practice?

Do the Math

We now move from first graders who were thinking about the relationship between addition and subtraction to a class of third graders working on the relationship between multiplication and division. In preparation for viewing the video, try this math for yourself.

1. Make a drawing or diagram or build a physical model for each of the following two problems. Write an equation for each.

   a. There are 5 baskets of crayons in the room. Each basket has 7 crayons. How many crayons are in the room?

   b. There are 35 crayons. Each basket holds 7 crayons. How many baskets hold crayons?

2. Can you draw or build a single representation that shows both problems? Explain to a colleague how your representation shows both problems.

(Continued)

(Continued)

3. These two problems are an example of a big idea about the relationship between multiplication and division. Can you write an "if . . . , then . . ." sentence that expresses the general relationship between these two problems?

   If . . . , then. . . .

   Can you use your representation from question #2 to explain why your "if . . . , then . . ." sentence is true?

4. Share your "if . . . , then . . . " sentence with a colleague. What is the same about your sentences, and what is different? How do your representations demonstrate why your statements are true?

Watch the Video: "Division Is Like Reversed Multiplication"

The video clip you'll be viewing is from Jeff Parks's third-grade class. Mr. Parks writes, "This video was taken on the third day after I returned from six weeks of paternity leave, so we were in the process of relearning our discussion norms. During this discussion and throughout the lesson, I knew I needed to reinforce listening to each other, because it had been so long since we had a whole-class discussion." Given the short timeline, many students had not been able to return their consent forms in time to be in the video that day and had to participate off camera; that is why you will see only about half the class in the video clip. Note that all students, including those off camera, participated in the class session, but the short clip you will view includes only students with consent to be videotaped.

A focus of Grade 3 is to understand the actions and behaviors of multiplication and division, how they are different from addition and subtraction, and how they can be used to model different problem contexts. For this lesson, students had worked individually on four related story problems:

1. There are 5 classrooms in third grade. The principal gave 7 books to each of them. How many books did the principal give to the third-grade classrooms?

2. 35 new books came to the school. The principal shared them equally among 5 classrooms. How many books did each classroom get? Did the first problem help you solve this problem? How?

3. There are 5 baskets of crayons in the room. Each basket has 7 crayons. How many crayons are in the room? Did the first problem help you solve this problem? How?

4. There are 35 crayons. Each basket holds 7 crayons. How many baskets have crayons? Did the other problems help you solve this problem? How?

Students created a representation for each of the problems (Figure 3.2). The point of this work wasn't to solve the problems—most students in the class knew the answers to 7×5, 5×7, $35 \div 5$, and $35 \div 7$. Or, if they didn't as they started this work, they knew them very well by the time they got to the discussion. Rather, the focus of the discussion you're about to see is to use the representations to think about the relationship between multiplication and division.

Figure 3.2 • One Third Grader's Work on the Four Story Problems

For each problem, write an equation, draw a representation, and write your solution.

1. There are 5 classrooms in third grade. The principal gave 7 books to each of them. How many books did the principal give to the third grade classrooms? Equation: $7 \times 5 = 35$ $7 + 7 + 7 + 7 + 7 = 35$ $7, 14, 21, 28, 35$	3. There are 5 baskets of crayons in the room. Each basket has 7 crayons. How many crayons are in the room? Equation: $5 \times 7 = 35$ 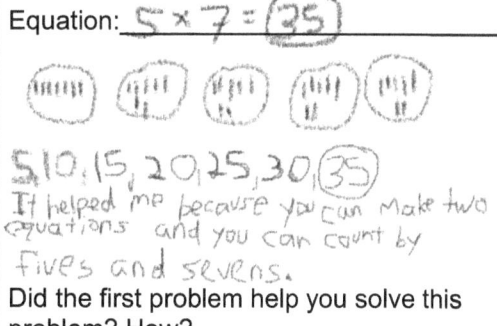 $5, 10, 15, 20, 25, 30, 35$ It helped me because you can make two equations and you can count by fives and sevens. Did the first problem help you solve this problem? How?
2. 35 new books came to the school. The principal shared them equally among 5 classrooms. How many books did each class get? Equation: $35 \div 5 = 7$  Did the first problem help you solve this problem? How?	4. There are 35 crayons. Each basket holds 7 crayons. How many baskets have crayons? Equation: $35 \div 7 = 5$ 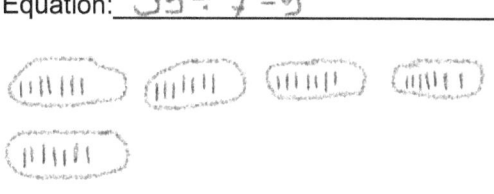 Did the other problems help you solve this problem? How?

For the class discussion, Mr. Parks selected three examples of students' work to share. As they discussed each representation, students considered whether it showed only multiplication or whether it could show division as well. Before the portion of the discussion you will view, one student, Duyen, had offered, "Division is like reversed multiplication." At that point, Mr. Parks asked the students to talk with a partner to consider this idea.

We will view the last four minutes of the discussion, just after the turn-and-talk about Duyen's observation. Mr. Parks has sketched a version of one of the student's representations (Figure 3.3) on the whiteboard to consider as they talk. We join the class as he asks the students what they were talking about during the turn-and-talk.

Figure 3.3 • A Student's Representation of 35 Crayons Divided Equally Into 5 Baskets

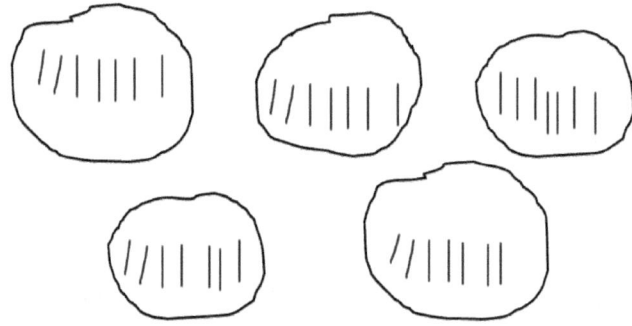

First Viewing of the Video: The Mathematics Students Are Working On

Watch Video 3.2, "Division Is Like Reversed Multiplication," with a focus on the mathematics content students are learning.

Video 3.2

"Division Is Like Reversed Multiplication"

qrs.ly/grfs4vk

Reflecting on the Video:
The Mathematics Students Are Working On

[You may want to use the transcript at the end of this chapter as you consider these questions.]

1. What are the important mathematical ideas in this clip?

2. How are students engaging with these ideas? What different ideas are students working on? Are there different openings into the mathematics?

3. In what ways is the mathematics challenging and engaging for the students?

4. What does the teacher do to focus the discussion and to promote persistence with complex ideas?

Second Viewing of the Video:
Making Space and Time for Every Student

Now rewatch the video clip, "Division Is Like Reversed Multiplication," with a focus on how different students participate in the discussion.

Reflecting on the Video:
Making Space and Time for Every Student

[You may want to use the transcript at the end of this chapter as you consider these questions.]

1. What are different ways that students participate in this discussion? How does the teacher make space and time for different students to contribute?

2. Look back at the contributions of Kenny, Aleeyah, Zahria, and Brielle. How does the teacher make space and time for different students' voices during this discussion? Here are some specific questions that might be useful in thinking about this:

   a. Kenny articulates quite a complete idea about multiplication and division at the beginning of this segment. Why do you think Mr. Parks doesn't reiterate that idea and wrap up the discussion at that point? How does Mr. Parks use Kenny's contribution to continue the class discussion?

   (Continued)

(Continued)

    b. Why might Mr. Parks ask Aleeyah to repeat her idea? In what ways does his response to what Aleeyah says invite other students into the conversation?

    c. How does Mr. Parks react to Zahria's hesitation, and what is the result?

    d. What do you notice about the interaction between Mr. Parks and Brielle?

3. If you were the teacher in this classroom reflecting on this lesson, what might you want to make note of in order to strengthen student participation? Are there aspects of the lesson that worked well to invite students into the mathematics? Are there questions you have about how you could better encourage students' voices and help students develop agency as mathematicians?

Read and Reflect on What Others See in the Video

What does it mean to make time and space for every student? How does a teacher keep in mind all the varying needs and experiences of each student and consider what could be an important next step for each of them? For one student, standing up in front of the group and pointing to something in another student's displayed representation, without saying a word, might be a huge step forward. For another student, repeating in their own words what another student has said might be an important contribution.

In the examples in this book, elementary students are learning to articulate deep and complex ideas about the operations. It's to be expected that many students will start to say something, then stop, backtrack, start again, become tongue-tied and break off, as Zahria did, or in some other way try to withdraw when they become unsure of what they intended to say. So often, the idea is there, but finding the words to explain it is difficult. It may seem kind to acquiesce to a student's withdrawal by giving them an out, for example, "Should we come back to you?" or "Do you want to call on someone else?" And, indeed, there may be times when this strategy is appropriate. But moments of hesitation can also be openings to communicate to students how important their ideas are. Moving too quickly to accept a student's reluctance can send the message that their ideas are not worth waiting for.

By repeating part of what the student said, asking them to start again, stating that it sounds like they're onto something important, or asking them to refer to a representation to show their idea, the teacher communicates curiosity and provides another opportunity for the student to express their idea.

1. The Teacher Reflects: Defining Goals for Individual Students

After viewing this video of his own teaching, Mr. Parks wrote about three students in the class, providing a window into his thinking about how to give space and time to each of them. We have given the students pseudonyms, but otherwise the accounts are as Mr. Parks wrote them. As you read his remarks, consider what Mr. Parks thinks about the participation of individual students.

Mr. Parks: *Canella.* Socially Canella is outgoing and positive. She easily makes friends and always has something to share with them. I can get her talking during small group reading assignments, but she tends not to participate in math class. In the regular math block, I noticed that she seemed distracted or unfocused, often when she was presented with multiple problems to solve. It wasn't until we got deeper into the lessons about investigating generalizations that I saw Canella come alive with really thoughtful ideas and explanations and creative representations that no other students were doing. What I learned about Canella was that she needed the time and space to wrestle with an idea in different ways. My guess is that during math class, the ideas had never seemed to solidify because she felt rushed by the multiple problems presented to her or the constraints of the activity. The openness of noticing and conjecturing, and the opportunity to go deeper with the same idea, allowed Canella to be creative in her understanding and understand this work in her own way.

However, during whole-group discussions, Canella still seemed distant or removed. In December, during some work on the commutative property of multiplication, Canella developed a representation that showed how you could turn an array of 6 × 4 ninety degrees to show 4 × 6. Because she was reluctant to speak in front of the group in math, I took a video of her explaining her representation and showed it to the class during a whole-class discussion. I wanted her to know she was onto something important and feel confident in her work. Sharing the video allowed us to hear her thoughts without putting the pressure on her to share in the moment. Figure 3.4 is her representation.

Figure 3.4 • Canella's Work to Show Why 6 × 4 = 4 × 6

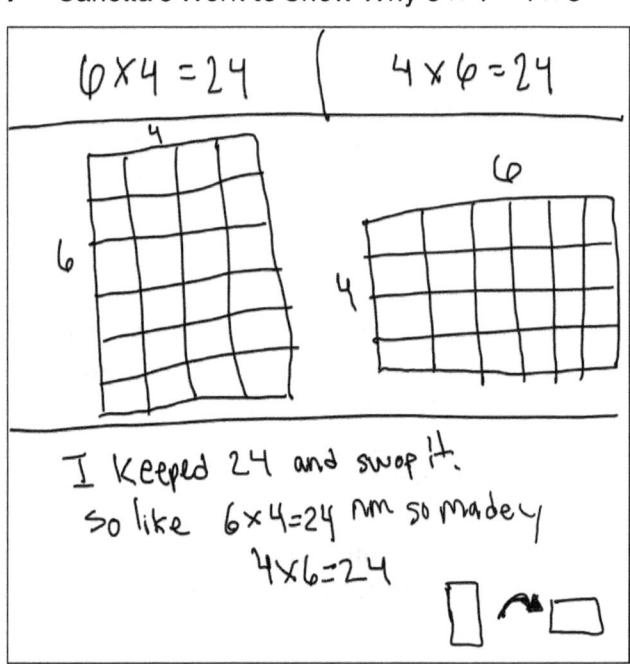

Being able to showcase her work in a subject that has not always been her strength had Canella feeling excited and valued as a thinker. After this moment, Canella's engagement in math lessons increased. In the lesson sequence about noticing and conjecturing about multiplication and division, Canella is engaged with the discussion by listening to her classmates and working to put the ideas into her own words. These lessons were the first times during a whole-group math discussion that Canella actively participated and was able to take the time to process her thoughts in front of the class.

· · · · · · · · · · · · · · ·

Darren. Darren was new to the school this year and came into my class with a history of challenges in school both academically and behaviorally. He did not feel confident in his ability as a learner and was reluctant to participate or share his ideas during class discussion. He would get frustrated by work that he found challenging, often leading to tears, would avoid work by asking to go to the bathroom repeatedly, or he would sit in silence staring at the ground. During our sessions on investigating generalizations, Darren began to build confidence in his ideas because this work gave him access to participate with his classmates while understanding that math is about making sense and not about having the right answer all the time.

One big concept that Darren has been working on throughout the year is that multiplication works differently than addition or subtraction. When seeing any word problem, he would create a picture that would show two sets of objects added or subtracted together. Our focus on creating representations helped him begin to think about multiplication as a number of equal groups. The following picture is Darren's representation of a multiplication story problem earlier in the year. He had erased his original picture which showed 7 boxes minus 3 boxes and had redrawn it as 7 groups with 3 in each (Figure 3.5).

Figure 3.5 • Darren's Representations of 7 Groups of 3

In the beginning of the lesson about the relationship between multiplication and division, Darren excitedly raised his hand to share what he and his partner discussed after I posed the question, "What do you notice about how Thu represents each of these problems with pictures and equations?" He said, "She was skip counting, and she did the circle and the tally marks," which starts the conversation about how all multiplication and division problems can be represented with groups and the same number in each. He continued to stay engaged throughout the lesson and also created his own representation of 6×3 as six circles with 3 lines in each.

This continues to be a process for Darren, but his ability to create representations has improved, and he has become more comfortable making mistakes and putting in the effort to solve problems that he may not understand right away. He has begun to offer his ideas in whole-group discussion and has even stood in front of the class to draw a representation for his classmates to work from. Darren has begun to feel confident in himself as a learner and is willing to put in the effort to understand new concepts that he would have given up on earlier in the year.

• • • • • • • • • • • • • •

Makayla. Makayla is a confident learner and has always been interested in sharing her ideas in class. She is accurate with computation and able to apply to her work the strategies shown in class. She enjoys solving problems and knowing she has the correct answer. When we began working on the lesson sequences about exploring generalizations, Makayla was resistant to diving deeply into concepts and would often claim that she was finished with the assignment. I had to question her regularly to encourage her to be exact in her representation and explain her thinking.

Throughout the year, I have been working with Makayla to help her make her representations clear and use precise language to explain them. As time went on, Makayla became more comfortable with expanding on her thinking and seeing connections. She began to see math as a time to develop ideas instead of only supplying a stagnant right or wrong response. She began to ask more questions and tackle new problems in different ways. During our discussion of the relationship between multiplication and division, Makayla was working hard to process the idea for herself, and her comments during class indicated to me that she was experiencing a spark of realization.

Shortly after the lessons on multiplication and division, our school closed because of the Covid outbreak, and I continued meeting with students online. Even working remotely, Makayla pushed her clarity of language and understanding. She submitted the following model and conjecture she had created on the computer (Figure 3.6).

Figure 3.6 • Makayla's Representation Illustrates Both $4 \times 5 = 20$ and $20 \div 5 = 4$

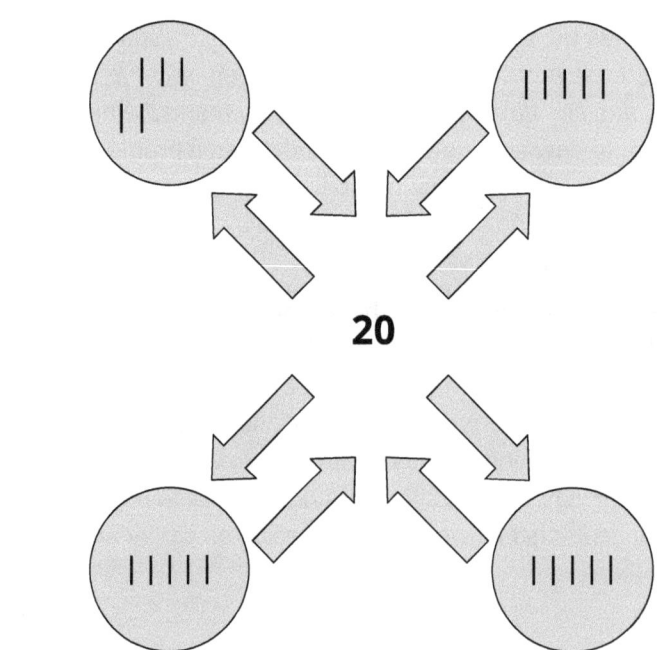

I have been doing lesson sequences on generalizations about how the arithmetic operations work for three years now, and with each group of students there are many stories like these. This focus on the behavior of the operations highlights real mathematical thinking and provides opportunities for all students to grow at their own pace. In this lesson (shown in the Video 3.2), all students were working on the same task, but as illustrated by Canella, Darren, and Makayla, each had a different experience with it. Through these lesson sequences, Canella learned that math is about creativity and collaboration. Darren learned that through clear representations he can solve problems he was too nervous to attempt before. Makayla learned that the depth behind mathematical concepts is thought provoking and worth exploring. As I continue this work, I am excited to see where this shared learning experience takes my students and me next.

2. Beliefs That Underlie the Teacher's Thinking: "If I Only Present One Way and You Don't Understand It, That's Not Equitable."

Later, in an interview, Mr. Parks reflected on the beliefs that underlie his approach:

You have content standards. I want students to leave my class knowing certain content. But because of the differences in their strengths, needs, and levels of comfort, all students come in at a different spot. By centering math around the thinking and the process of learning, yes, I'm teaching this content, but I'm focusing on mathematical practices. If students are building mathematical practices, they're building skills they'll transfer throughout their learning career. To allow equitable participation, they're engaging with the same content, but they're getting what they need from the process. And they're not feeling like this is above them or beyond them. One of my norms is that everyone can learn math, but the paths they take to get there are going to be different. I'm trying to open up the thinking and open up the task and the work without teaching one way to do it. If I only present one way and you don't understand it, that's not equitable. If it doesn't work for you, that's not equitable. As the teacher, I'm looking for what individual students need to get there, but centering it around how the students work together.

Reflecting on Mr. Parks's Writing

Consider again the main ideas of this chapter as you reflect on Mr. Parks's writing.

1. What are some of the issues Mr. Parks takes into account as he thinks about the participation of Canella, Darren, and Makayla? How does he think about both math learning goals and goals for participation for each of the students?

2. What are key beliefs that underlie Mr. Parks's practice? How are your own beliefs similar or different?

3. Consider Mr. Parks's statement, *"One of my norms is that everyone can learn math, but the path they take to get there is going to be different."* How do you respond to this statement? Do you assume that every student in your class can learn math? How do you see this occurring or not in your own context? Do you believe every student has mathematical ideas worthy of being listened to and shared?

What Do You Want to Remember From This Chapter?

In this chapter, Mr. Parks shared some of his thinking as he attended to the development of a mathematics community in his classroom. Before leaving this chapter, think about attending to both the needs of the group as a whole and the individuals in it as you build mathematics community in your own context. Here are some explicit moves we want to make sure you keep thinking about.

- **Practice patience.** Both teacher and students are learning the patience and concentration required to attend to fledgling mathematicians' attempts to articulate their mathematical ideas. Pauses, backtracking, hesitations, trailing off, and starting over are all expected parts of expressing ideas. Expect and encourage all efforts.

- **Normalize uncertainty**. When a student starts to speak in whole-class discussion and then withdraws or hesitates, treat this as a normal occurrence and find ways to encourage them to continue with their idea.

- **Emphasize mathematical practices**. Mathematical practices such as creating clear representations of a problem, looking for connections across representations, explaining one's thinking, and analyzing the thinking of others are skills students will carry throughout their learning careers.

- **Consider goals for each student.** At the same time that a teacher has a class goal for a lesson, the teacher will also have different goals for different students depending on their individual strengths, challenges, and knowledge.

Taking a Next Step

Think of one or two students who do not typically volunteer to contribute ideas to class math discussions. (You might want to refer back to the list you made at the end of Chapter 1. Choose someone different from the student you selected for the Next Step in Chapter 2.) For the next week, make a point of observing and talking with them about their work during math time. Make note of what they say, how they solve problems, what representations they create. Think about how their thinking might be shared with the class. Determine a strategy for including the one or two students you've focused on in some way in an upcoming math discussion. What preparation do you need to do with the student(s) ahead of time? If you don't have your own classroom, adapt this activity to another group with which you are working.

Video 3.1 Transcript: **"I Mean Adding"**

Ms. Schooler:	Was there anything we noticed about the first problem that helped us with the second problem?
Ms. Schooler:	Jeuri, talk to us!
Jeuri:	So . . . [20-second pause]
Ms. Schooler:	Jeuri, what was similar about the first problem and the second problem? Or what was different?
Ms. Schooler:	Hands are down. We're letting Jeuri think and talk.
Jeuri:	It's minus 5 and then it was 7 left so it was . . . [28-second pause]
Ms. Schooler:	You got this, Jeuri.
Another student:	We believe in you.
Jeuri:	12 minus 5 equals 7.
Ms. Schooler:	Now what's the same about that as what was in the first problem, Jeuri?
Jeuri:	You mean this? [points to first problem on the chart]
Ms. Schooler:	Yes!
Jeuri:	Oh. So it was 7 and then it makes it together [shows hands apart then claps hands together].
Ms. Schooler:	So the first problem is together? What do you mean?
Jeuri:	I mean adding.

Video 3.2 Transcript: **"Division Is Like Reversed Multiplication"**

Kenny:	I think that if you do the 7 times 5 and it equals 35, the division is basically like reversed because the answer for 7 times 5 is the thing you have to divide for, split up. So then if we do like 35 divided by 7, it has to be equal 5.
Mr. Parks:	So all this, we said, when we did, as David mentioned, 5 groups of 7 gives us what? [Students: 35] 35. Is everyone with me on that? So let's just say all this together, this is our 35. And then Kenny said something interesting about what happens when we divide it. What happens when we take— now are dividing with this 35?
Aleeyah:	I want to add on to Kenny. So when you're dividing the 35, when you divide it, you are putting the 35 inside of groups. And then with the multiplication, you are putting the groups together to get 35.
Mr. Parks:	Could someone repeat that? [pause] Maybe say that one more time, Aleeyah.
Aleeyah:	So with division you are putting the 35 into groups, but with multiplication you are putting the groups together to make 35.
Mr. Parks:	How can I show that with this representation, do you think? How can I show that . . .
Zahria:	You put, you put all of them together and then, like, I don't know . . .
Mr. Parks:	What do you mean, "put all of them together"?
Zahria:	Cause, like, you put, you put all of them together to get 35, and then you split them apart to get, to do division, you split them all apart, and for multiplication, you put them all together.
Mr. Parks:	Can someone repeat that? I think we're getting to this idea of why it's reversed, this idea of putting together and what else? Brielle?
Brielle:	You said to repeat her? [Mr. Parks: Yeah.] She said, if you like get the, I don't know how to say it, if you like put them all together, then it will make the number.
Mr. Parks:	What number?
Brielle:	35.

Mr. Parks:	So for multiplication, we're putting together. And what happens when we reverse it?
Several students:	It goes back to times . . . you're taking . . . you're like . . .
Mr. Parks:	Jadyn, what happens when we reverse it? This idea that it's reverse multiplication.
Jadyn:	You're taking it back apart. It's like you're tearing it apart.
Zahria:	You're putting it back in groups.
Aleeyah:	You're breaking it apart.
Jadyn:	It's like you are tearing it apart.
Kenny:	Yeah.
Mr. Parks:	Interesting. So one we're putting it together, okay, and then the next one we're tearing it apart, breaking it apart.
Mr. Parks:	We're going to keep with these ideas, and I'll try to do a better job of charting it and making it pretty, but we're going to think about how we can kind of create these representations that show this action, and make a clear statement or conjecture about this relationship. And I think this is going to really help us see these patterns. Final thoughts?
Aleeyah:	It's actually like when . . . it's like subtracting and adding.
Mr. Parks:	Interesting. Ah! What time is it? [laughs] Can we just leave that and come back to it because of time's sake? But interesting—it's a similar comparison, as Aleeyah said, to subtraction and addition. Hmmm. . . . Thank you, everyone, for kind of participating and sharing your thinking around that. We'll come back after I make a nice pretty poster and keep working on this idea.

Chapter 4

Encouraging Persistence

Think back to the video of Jeff Parks's third graders in Chapter 3. At one point in the class discussion, Zahria starts expressing her thought about the relationship between multiplication and division, then trails off: "You put, you put all of them together and then, like, I don't know . . . " Mr. Parks might have accepted her reluctance to continue by using common teacher moves like, "Should I come back to you?" or "Do you want to call on a friend?" But instead, he looks for the sense in what she is beginning to articulate, and repeats her own sense-making back to her, "What do you mean, 'put all of them together?'" He *normalizes her hesitation*, while *demonstrating his curiosity* about her idea. But his question also does something else. It encourages Zahria's development of her own agency as a learner of mathematics. Rather than turning to another student, he communicates his expectation that she can persist in formulating and expressing her own ideas.

When we facilitate math discussions that are about complex ideas, expressing those ideas takes work. Further, the act of verbalizing ideas is itself an act of learning. As we see in many of the videos in this book, students become clearer about their own ideas in the process of trying to express them—by repeating, backtracking, rewording. Students need to learn that persistence is necessary when working on deep and rigorous mathematics. If we want every student to learn that they have mathematical ideas worth sharing, we need to provide experiences in which they have the time and encouragement to persist.

In this chapter, you will

- do some math about equivalent addition expressions to prepare for watching a second grader working on this idea,

- view a video of a teacher working with one student during pair work as the student formulates an idea, and

- consider commentaries by our Critical Friends and by the teacher, Ms. Clarke, in order to reflect on the themes of this chapter.

Many students need opportunities to formulate and articulate their ideas before they are ready to share them in a whole-class setting. In Chapter 2, we considered the role of turn-and-talk as one structure that supports students to develop their own ideas in the midst of a whole-class discussion. In this chapter, we look at an example of students working in pairs for a more extended period of time and how the teacher interacts with a student during pair work. As you consider the classroom video and the commentaries in this chapter, think about the following aspects of the principle that *every voice matters* and how they support the interweaving of rigorous mathematics and equitable participation.

1. When articulating new ideas, students need time to repeat themselves, backtrack, and reword as they're formulating and expressing their ideas.

2. When students seem stuck, look for the beginnings of important ideas in what they've said, and reflect these kernels back to them in their own words.

3. When a student has offered an important idea, consider reflecting it back to them, using additional vocabulary or math terms that could be useful for them in making their ideas clear.

Do the Math

The second graders in the video you'll be watching are working on pairs of addition expressions that have the same sum. In order to get a sense of how the students are thinking about these examples and beginning to develop deeper insights about the structure of addition, try these math activities.

1. Take a look at these examples. What do you notice?

   $10 + 5 = 9 + 6$

   $18 + 18 = 19 + 17$

   $8 + 12 = 9 + 11 = 10 + 10$

2. These equations are examples of something that is true about addition. Write a general statement that describes what you noticed. You might use this format: "If . . . , then . . .

   Compare your statement with a colleague's statement. Notice how the statements are similar or different.

Watch the Video:
"I Don't Know How to Explain It"

Quayisha Clarke's second-grade class has been working on equivalent addition expressions. In a previous class, they created a list of expressions equivalent to 15, in order: $0 + 15$, $1 + 14$, $2 + 13$, $3 + 12$, and so forth, down to $15 + 0$. In another session, they came up with equations like the ones you just worked with that show the equality of two related addition expressions. Ms. Clarke begins the class by drawing students' attention to a poster of these equations (see Figure 4.1).

Figure 4.1 • A Poster of Equations Generated by Ms. Clarke's Students

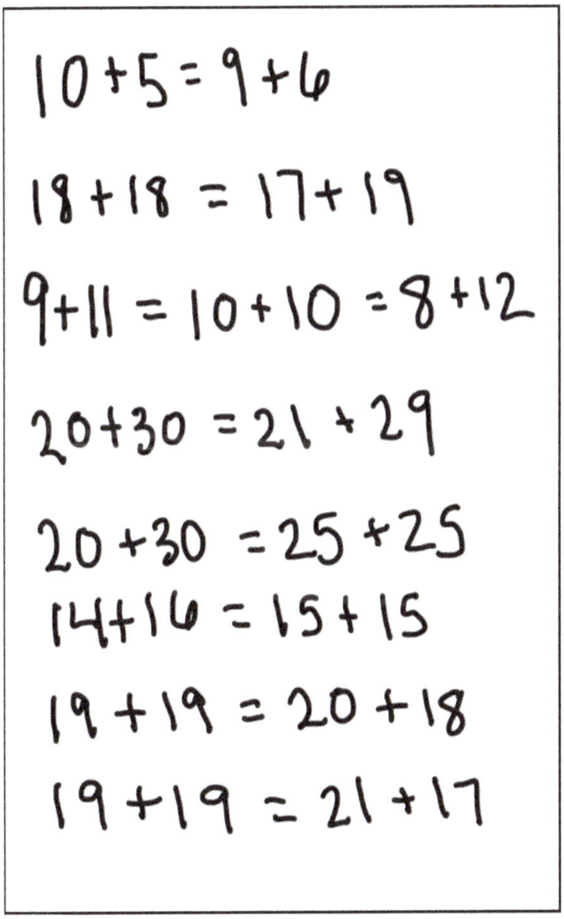

$$10 + 5 = 9 + 6$$
$$18 + 18 = 17 + 19$$
$$9 + 11 = 10 + 10 = 8 + 12$$
$$20 + 30 = 21 + 29$$
$$20 + 30 = 25 + 25$$
$$14 + 16 = 15 + 15$$
$$19 + 19 = 20 + 18$$
$$19 + 19 = 21 + 17$$

Ms. Clarke suggests students focus on one of the equations and poses the question, "How do you know these expressions are equal? . . . How can we explain, maybe to a first grader, what's happening here?" She sends students off to work in pairs. Each pair has a clipboard and a student sheet that shows some of the equations and gives them space to write their explanation. As students work in pairs, Ms. Clarke circulates, checking in with each pair. In this clip, she is meeting with Leo and Kingston but is working primarily with Leo as he tries to articulate his idea. You will read her comments later about why she was focusing on Leo during this interaction.

First Viewing of the Video: "I Don't Know How to Explain It"

Watch Video 4.1, "I Don't Know How to Explain It," with a focus on the math ideas Leo is working on and how Ms. Clarke is supporting his mathematical thinking.

Video 4.1

"I Don't Know How to Explain It"

qrs.ly/p1fs4vq

Reflecting on the Video: The Mathematics Students Are Working On

[You may want to use the transcript at the end of this chapter as you consider these questions.]

1. What is the mathematics content that Leo is working on?

2. Leo already knows how to add $10 + 5$ and $9 + 6$ and also already knows that $10 + 5$ and $9 + 6$ have the same sum. What is it, then, that is challenging for him? Why do you think it is challenging?

Second Viewing of the Video: "I Don't Know How to Explain It"

Now rewatch the video clip, "I Don't Know How to Explain It," with a focus on what you notice about Leo's persistence and how Ms. Clarke supports his efforts to articulate his ideas.

Reflecting on the Video:
Encouraging Persistence

[You may want to use the transcript at the end of this chapter as you consider these questions.]

1. Identify and describe the specific moves Ms. Clarke makes to encourage Leo to persist in thinking through and articulating his idea. What results from the choices she makes?

2. Later, when the class comes back together to share some of the work they have done in pairs, Ms. Clarke says to the class, "This is really hard to put into words. It's a little challenging to talk about what's happening to these expressions." What might be important about her choice to describe their work in this way?

Read and Reflect on
What Others See in the Video

Let's return to the three aspects of the principle that *every voice matters* listed at the beginning of this chapter in order to reflect further on the interaction between Leo and Ms. Clarke:

1. When articulating new ideas, students need time to repeat themselves, backtrack, and reword as they're formulating and expressing their ideas.

2. When students seem stuck, look for the beginnings of important ideas in what they've said, and reflect these kernels back to them in their own words.

3. When a student has offered an important idea, consider reflecting it back to them, using additional vocabulary or math terms that could be useful for them in making their ideas clear.

When our Critical Friends viewed "I Don't Know How to Explain It," their comments centered around four ideas—encouraging persistence, Leo's style of discourse, using small group work to understand student thinking and plan for later discussion, and Kingston's participation.

1. The Teacher's Moves to Encourage Persistence

Virginia Bastable: I wanted to talk about what Ms. Clarke does when Leo is having trouble expressing his thinking and says, "I don't know how to explain it." She reports back his exact words. And then, when he is still hesitant, she gives him a suggestion—do you want to use a specific number? And then he continues. The part I like about this clip is that it shows how hard it is to listen to kids when they're not completely clear and how we don't often give them enough time to let them be messy because we're uncomfortable on their behalf.

Lynne Godfrey: I almost thought she was going to say, "Do you want to phone a friend?" And she didn't do that, right? She didn't take it away from him but was thinking about how to help him take this up. For example, she suggested a more specific case instead of the more general question about what's happening. There are two things that I feel are really important that I see students learning in this clip. One is that math class should be a place where you can say, "I don't know." And the second is that in math class you can express your frustration when you can't find the words to say your idea. It's not about speed or about who gets the right answer but that math class is hard, right? So, I love seeing Leo say "I don't know" and be able to express his frustration.

Darlene Ratliff: [When Kingston was pointing at the worksheet], indicating "It's right here," It was as if Leo was thinking, "I want to do it myself. I want to process this myself. I won't just read this and still not understand it." He was responsible for his own learning, knowing that the teacher supported his process. I think she used his own words, which gave him even more strength to continue to process. She heard him, and he knew that by her response. She didn't just say, "Yep, okay, all right," no, she gave what he was saying right back to him.

2. Supporting Leo's Own Way of Imagining the Problem

Cindy Ballenger: Ms. Clarke says, "So the 10 decreases to 9 and the 5 increases to 6?" And he says, "Yeah." You might think, oh, she's given him a word he needed. But then he says, "It's like the 10 shared one of his numbers with the 5," and he goes into his own thing, and she lets him do it.

Marta Garcia: I love that Leo is making a story about someone taking action, someone who's coming and making that change. They're taking it, they're giving it, and they're sharing. He's trying to articulate what's happening by putting a story around it where something's getting moved and shared.

Cindy Ballenger: First he says, "Somebody takes away 1 from the 10." And then the next time he has the numbers being the actors and they want to share. I'm not sure which I like better. I love them both. I think I like the numbers sharing the best.

Virginia Bastable: It felt very personal, you know, it was like the way you want to treat people—the numbers should be treating each other that way.

3. Using Small Group Work to Understand Student Thinking and Plan for Later Discussions

Lynne Godfrey: It's really important for teachers to listen in on what's happening in the small group, who's doing the work, and what questions are coming up in the group. Sometimes whoever holds the clipboard and the pen rules and is the speaker, and that was not the case in this episode. I think that's often the case, especially when kids go off to work on their own.

Hetal Patel: In this case the teacher lingers with a student to get a better understanding of what the student's ideas are. Through that she gathers information about what they're learning and what they're thinking. That influences how the teacher facilitates the whole group and how to make so many different micro decisions you may need to make for the whole group.

Marta Garcia: And I'm also thinking about the role of oral rehearsing, especially if you have students who might not have the language or who might need time to consolidate their ideas. By spending that time, that one student has had time to oral-rehearse. This type of support is especially important for multilingual students. Encouraging students to express their nascent ideas in their primary language and supporting the use of translanguaging (moving fluidly between languages), both in small groups and in the whole group, positions multilingual learners as competent and able to contribute to the mathematical ideas of the class. Oral rehearsing gives students a chance to try out drafty ideas. Then later on the teacher can say, "Would you like to share your ideas?"

Cindy Ballenger: Leo had the opportunity to state his idea and then repeat that same idea—to state it using different words. As Marta said, that's important for multilingual learners and students for whom academic vocabulary is less comfortable. Repeating provides students the chance to revise as they are restating their thought. They are thinking through whether they agree or disagree with their own original articulation. As they hear their own repeated voicing of ideas, they might also be thinking through grammar and vocabulary, changing their words a little each time. While the student is restating, teachers have an opportunity to think about what to ask or to jot notes about what the student is saying which they can reflect on later.

4. What About Leo's Partner, Kingston?

Marta Garcia: Leo's partner is never really invited into the conversation. He taps the clipboard and he's listening. I'm wondering what he has to offer and why he doesn't engage in the conversation or isn't invited into the conversation. So I think that's a question to think about.

Hetal Patel: And that's kind of what I was thinking, too. I was trying to speculate about the range of possibilities. Is it that Kingston is the one who gets it, maybe he's the one who did all the work and wrote it up? He's the one who has the clipboard. Darlene is always bringing up that there are so many nonverbal ways to participate. Is that part of what he may be doing? We don't know, but I'm thinking about what the different reasons might be.

Reflecting on Our Critical Friends' Commentaries

1. Hetal Patel talks about "lingering" with a student. In your own context, do you or other teachers spend focused time with individual students? When and why do you do that? What are implications for your own practice?

2. Lynne Godfrey mentions the teacher move we've all seen, and probably used ourselves, of asking a child who seems stuck to "phone a friend." What are your thoughts about using this move?

3. In what ways does Leo's explanation, and the words he chooses, indicate his agency as a mathematics learner?

4. How do you use or might you use oral rehearsal to help students prepare for sharing their ideas? Are there students in your context, including multilingual learners, who could benefit from the opportunity to oral-rehearse?

5. What are you thinking about why Ms. Clarke might not have involved Kingston during this interaction? In a similar situation, what are options you might consider?

The Teacher Reflects on Her Interaction With Leo

In this interaction between Leo and Ms. Clarke, she makes many decisions about how to support his ideas and encourage him to persist. What can we learn from the moves she makes? Three aspects of what she does can help all of us reflect on our own practice. First, she does not accept Leo's repeated assertion that he doesn't know how to articulate his thoughts. Second, she listens hard and identifies what he has said that he might go back to ("So you were saying that you take 1 from the 10"). By reiterating his own words, she both provides a way back into his thinking and acknowledges that she has heard something important in what he has started to say. A little later, she also acknowledges his own language of "shared" and "became a bigger number." Third, once he has more fully articulated what he wants to say, she repeats his idea back to him, while also providing some additional vocabulary, "decreases" and "increases," which might be useful to him as he continues his work. She is not insisting he use that vocabulary now but offers it for him to hear and connect to his own words.

Teachers might make different decisions for a different student, although the principle of *both* reflecting back a student's own ideas *and* judiciously providing additional vocabulary as it seems useful is important to keep in mind for all students. When and how to take these actions depends, in part, on what a teacher knows about each student.

As you have read, our Critical Friends watching this video clip also noticed that, in this short interaction, Ms. Clarke only talks with Leo, even though Kingston, his partner, is urgently indicating the worksheet they had been working on together. Why might she be lingering with Leo and, at least for these two minutes, not including Kingston?

In an interview about the video clip with Ms. Clarke sometime later, she touched on these issues.

Ms. Clarke: I want kids to be precise in their language and to explain what's happening or what they're noticing and then try to see if they can notice it in more than one problem. That's a big goal: How are these problems related, and why do they all work in the same way? And I felt like if Leo could understand what's happening for 10 plus 5 and 9 plus 6, maybe then he could use the same reasoning with a different problem. But he first had to be able to explain it in his own words.

So I tried to facilitate Leo's thinking. I wasn't telling him what to think or how to think. I was providing choices to uncover the math but still allowing him his own entry. You can tell he was thinking. He was very much focused and wanted to work hard. I could see that he was struggling with just trying to get into words what he was thinking. I wanted to provide support without leading too much.

[When I watched this video again,] I was wondering why I didn't include Kingston. Kingston was pointing to the paper and trying to show Leo. I think I wanted to focus on Leo's thinking because it looked like he was really working hard, and Kingston didn't chime in, so I didn't want to put him on the spot.

There's so much going on in all of the kids' thinking, and you're trying to make space for all of them. [Some students] feel comfortable first sharing an idea one-on-one or in a small group and after that sharing in the whole group because they've already said it and have gained confidence. Even if a student is not sure, the fact that they're wrestling with the idea is important. The interest and curiosity is a goal, too. You can alienate people from math if you're not careful or make them feel like they don't have a space, if it's the same kids always talking. Or you can boost up people's confidence in math and help them think, 'Well, there's many ways to think about things and to show things, there's lots to uncover here, and I have something to say.'

After students worked in pairs that day, Ms. Clarke gathered them in the whole group to share some of their ideas. While Leo wasn't one of the students who shared, he continued to refine his thinking as the class continued its work on equivalent addition expressions. In a subsequent lesson, the second graders are applying the ideas they've developed to solve a difficult addition problem. Ms. Clarke writes this expression horizontally on the board: 399 + 234. She says to the class, "I want you to think about an expression that would have the same sum and would be easier for you to solve. Give me a silent thumb when you have an idea. What would be an expression that has the same sum and would be easier to solve?" Both Kingston and Leo participate in the whole group. Kingston first proposes, "You can take 1 from 234 and add it to 399, so it can be an easier way—it's going to be 400 plus 233." Leo adds, "400 + 233 is easier than 399 plus 234. It's easier because it would be hard to figure it out, and 400 plus 233 you can just tell 400 plus 200. That equals 600, and add the 33, it equals 633." Both students seem confident and eager to contribute to solving this more difficult problem.

 Reflecting on Ms. Clarke's Writing

1. Find a sentence in Ms. Clarke's writing that you'd like to think about further. Why is this sentence meaningful to you?

2. What is a question about your practice that Ms. Clarke's reflection brings up for you?

What Do You Want to Remember From This Chapter?

Take a few minutes to note for yourself ideas you want to hold onto as you continue to investigate the meaning of a mathematics community and how to build it. What teacher moves have you noticed in this chapter that you want to bring into your own practice? Here are some of the specific ways to think about how every student can have a voice in the mathematics classroom:

- **Encourage persistence.** There is an unfortunate myth about mathematics learning that you either know the correct response or you don't. Rather, help students learn that you are challenging them to think about hard ideas and that you expect them to need time and effort to make sense of them.

- **Embrace pauses, backtracking, silences.** In order to tolerate the repeating, backtracking, rewording, and rethinking that are *essential* parts of mathematical discourse, we have to come to terms with our own discomfort when students are uncertain, our own tendencies to want to save them from frustration, and our own difficulties in tolerating the time students take to articulate their thoughts.

- **Reflect back students' own words to them.** Help students see that they have kernels of mathematical ideas by reflecting back to them their own beginnings of sense-making. This can be as simple as "You started to say something about multiplication" or "You said something about a 1—where were you seeing a 1?" Reflecting back in this way indicates that you are listening hard and that you expect students to have ideas.

- **Give students opportunities to rehearse their ideas.** While you circulate during individual and small group work, engage students to explain their thinking. Identify ideas that could be shared with the whole group, keeping in mind which students often share and which do not. Ask more reluctant students in advance about how they might share their ideas, and give them opportunities to rehearse what they might show or say.

Taking a Next Step

· ·

Leo starts to articulate his idea, then, two different times, hesitates and says, "I don't know how to explain it." Write down five possible types of responses a teacher might make at this point. Draw on what you've heard teachers do or what you've done yourself when a student doesn't seem to be able to continue. Now look at your list, and consider the possible ways in which each response might affect a student's agency as a mathematics learner. Under what circumstances might you choose each response?

Video 4.1 Transcript: **"I Don't Know How to Explain It"**

Leo: Okay, for example, 10 plus 5 and 9 plus 6, they both equal the same sum.

Ms. Clarke: And how do you know that? How do you know that they both equal the same sum?

Leo: Ummm . . . [hesitates]

Ms. Clarke: What's happening there?

Leo: It's . . . [trails off]

Kingston: You can look right here [indicates what they've already written on the worksheet].

Leo: I think it's . . . I don't know how to explain it.

Ms. Clarke: OK, do you want to use the specific numbers [Leo: "yes"] or you can say "addends" . . .

Leo: The 10 and the 5 and, um, wait . . . Somebody takes away the 10, 1 from, somebody takes away 1 from the 10, and then they put it in the 8, and then it equals, then it makes 9, and the 6 was, wait . . . I don't know how to explain it.

Ms. Clarke: OK. So you were saying that you take 1 from the 10. And where does that go?

Leo: To the 9.

Ms. Clarke: It goes to the 9, or it becomes 9?

Leo: Uhhh, no, I think it goes to the 6.

Ms. Clarke: So the 10 decreases to 9 and the 5 increases to 6.

Leo: Uh-huh.

Ms. Clarke: OK. Look at $18 + 18$ . . .

Leo: Cause it's the 10, it's like the 10 shared one of his numbers with the 5 because the 5 wanted to become a bigger number.

Ms. Clarke: OK, so can we try to say that same, in your own words, you can say "shared" and "became a bigger number" with the other expressions here, what's happening? Look at $18 + 18 = 17 + 19$. How do you know those are equal?

Part Two

Collaboration Supports Student Agency

"I have an idea, and my partners have ideas, and together we have one big idea." In this statement, a third grader names an essential aspect of her math class, similar to what Aguirre and her colleagues (Aguirre et al., 2024) refer to as *collective mathematical agency.* This student recognizes that when she and her classmates collaborate, each sharing their thoughts and representations, they can dig deeply into mathematical ideas, and everyone's learning is enhanced. It's not simply a matter of exposing students to the "best" ideas the class can offer. Rather, one student's idea spurs another student's idea which spurs yet another. Even when a student offers just a kernel of an idea, even if the work contains an error or needs to be revised, the contribution can move learning forward. Making connections between what different students offer is new learning for everyone.

Historically, collaboration has often been discouraged in mathematics classrooms. If the end product of mathematics learning is simply the correct answer, and individuals are assessed on the number of their correct answers, collaboration can even be viewed as cheating. One of the authors remembers, from her own schooling, that gesture of curling her arm around her math paper so that no one else could see what she was writing. However, if the end product we are after is *making sense* of deep mathematics, collaboration is key.

As teachers, we are continually assessing what each individual student understands and needs to work on, so that we can plan next teaching steps. But that does not mean that each student must build mathematical ideas in isolation. Rather, the process of learning complex mathematical ideas benefits from enacting collective agency. Comparing different approaches, different representations, different ways of thinking, having one's own ideas considered and considering the ideas of others—these processes give students more tools and openings for making sense of mathematics.

In Part Two, we focus on how noticing patterns and developing conjectures based on those patterns engage students in building ideas collaboratively as they learn about the behavior and properties of the arithmetic operations. The mathematics of generalizing about the operations is significant for students' understanding of arithmetic as well as their later work in algebra. Students begin by looking at and describing patterns of something consistent happening across related examples. For example, when they are learning their multiplication facts, students notice the pattern that facts come in pairs, that $3 \times 6 = 6 \times 3$ and $7 \times 8 = 8 \times 7$, but they may not be asked to think about it any further. If given the opportunity, students can investigate more deeply what happens when you change the order of factors in a multiplication problem and why that happens. In Chapter 3, Mr. Parks describes how Canella shared with the class her image of rotating an array to show why $6 \times 4 = 4 \times 6$. Through studying examples of such a regularity, students begin to recognize the underlying structure of the mathematics, which they then try to put into words by articulating a mathematical conjecture—a statement that captures a pattern they have investigated and believe to be generally true: If you change the order of the factors in a multiplication expression, the product will be the same.

As they notice patterns and construct conjectures, students come to understand that they are considering difficult and challenging mathematics and that they can rely on each other to collaboratively build big mathematical concepts. Just as when discussing a book or a historical event, in mathematics, too, different students find different entry points and contribute their own, unique perspectives.

As students become experienced in the collaborative construction of mathematical ideas, they find their own voices, listen to others, expect to be listened to, and learn to value the interaction of their own and others' contributions to collective learning. They develop a stance of curiosity toward their classmates' thinking. Through collective investigation, students are also developing individual agency—learning that their own confidence and competence in mathematics doesn't mean they have to do everything alone but that collaboration increases every individual's mathematical power.

> **The following are the major themes of Part Two:**
>
> - Deep and complex mathematics requires collaborative construction of ideas.
> - Through collaboratively building complex ideas, students develop both collective and individual mathematical agency.
> - The mathematics of generalizing—in particular, noticing and conjecturing about patterns in the arithmetic operations—is an important site for collaborative learning about central instructional content.

Chapter 5

Noticing Patterns as a Gateway to Building Mathematical Ideas Together

As you've seen so far in this book, there are participation structures, such as turn-and-talk or ways to refer to previous comments ("I want to build on . . . "), that teachers use to encourage participation and collaboration. But there are also characteristics *of the mathematics itself* that can encourage collaboration. In this and the next two chapters, we consider two mathematical practices that encourage students to collaborate as they investigate complex mathematical ideas: (1) noticing patterns and (2) developing conjectures based on those patterns.

Patterns occur everywhere in every area of mathematics. In fact, mathematics has been described as the "science of patterns" (Hardy, 1967; Steen, 1990). Mathematicians attend to patterns, notice and describe the regularities they see, try to understand when and why they occur, then attempt to prove their generality and capture them in formal mathematical language and symbols.

In our lesson sequences, it is not simply for the sake of pattern-finding that we ask students to look for patterns. Rather, noticing patterns is a gateway, an important precursor to understanding mathematical structure. For example, in Chapter 4, Ms. Clarke's second graders had noticed that when adding two numbers, if one addend increases by 1 and the other addend decreases by 1, the sum stays the same. They were coming to see something important about how addition behaves. Noticing patterns draws students' attention to characteristics of the operation, and teachers can build on what students notice about the pattern to help them dig into important mathematical ideas.

Noticing is also a practice that invites students in. When students are asked, "What do you notice?" about a set of related expressions or equations, the floor is open for thinking and speculation. We have seen, again and again, in many classrooms in many contexts, how students' minds are captured and deeply engaged when they are invited to think about the underlying patterns of the mathematics.

In this chapter, you will

- investigate some patterns in subtraction for yourself, in order to prepare for considering students' thinking,
- view second graders noticing and describing patterns in subtraction, and
- consider commentaries by our Critical Friends and the teacher in the video about students' learning and collaboration.

Once students notice and describe a pattern, they naturally begin to try to articulate the essence of the pattern—to describe more succinctly what the pattern is, when it occurs, and why it holds. That is, they begin to construct conjectures. You will notice this happening toward the end of the video clip, and we will delve further into conjecturing in the next two chapters.

As you consider the classroom video and the commentaries in this chapter, we'd like you to think about four aspects of the principle that *collaboration supports student agency* in a mathematics community built on interweaving rigorous mathematics and equitable participation.

1. "What do you notice?" is an *invitation*, an opening for all students to enter into making sense of the underlying structures and behaviors of the operations.
2. As students notice and describe patterns, teachers are alerted to mathematical ideas that a range of students has to offer.
3. Noticing patterns in sets of related expressions leads naturally into developing conjectures about how an operation behaves.
4. Investigations into arithmetic structure, in which different students can work simultaneously on different aspects of the content, encourage collective and individual agency.

Do the Math

In Chapter 4, you viewed second graders working on equivalent addition expressions like $10 + 5 = 9 + 6$. In this chapter, you'll see students working on parallel ideas in subtraction. Thinking about subtraction is often, even for adults, more challenging than thinking about addition. Addition is the first operation that young students encounter. The idea of putting quantities together or adding on more to a quantity is familiar to them by the time they encounter other operations. When they begin to delve into subtraction or, later, multiplication, students often at first expect these operations to behave in the same way as addition. In order to orient your own thinking to the subtraction work you will see the students engaged in, try the following math task.

1. Look at each set of four expressions below. What do you notice?

SET A	SET B
$17 - 5$	$17 - 5$
$16 - 4$	$18 - 6$
$15 - 3$	$19 - 7$
$14 - 2$	$20 - 8$

2. What do you notice about Set A?

3. What do you notice about Set B?

4. Write out in words one or more statements about what you think is going on in Set A, in Set B, and/or in both. You might start your sentences with "If . . . " or "When . . . "

   If you are working with others, share your statements with your colleagues.

5. Look back at your math work from Chapter 4. What do you notice when you compare your work on equivalent subtraction expressions to your work on equivalent addition expressions?

Watch the Video: "I Want to Build on My Classmate's Thinking"

As you saw in Chapter 4, Quayisha Clarke's second-grade class worked on noticing and describing patterns in addition. They looked at equations such as $9 + 5 = 8 + 6$ and $31 + 18 = 30 + 19$, and in a later lesson, wrote out what they saw going on across many examples. They compared some of their statements and chose one of them that they thought conveyed clearly and precisely what they were noticing. Figure 5.1 shows some of the work that was recorded as the class put their thoughts together. In the course of that discussion, the teacher reminded students about the words *addend* and *sum,* so that this language would be available to students who felt comfortable using these terms. Students were welcome to use everyday language, math terms, or a mixture of the two. On the poster, the starred statement is what students chose as a class conjecture: *If you take away any number from one addend and add it to the other addend, the sum remains the same.*

Figure 5.1 • Ms. Clarke's Students' Work on Equivalent Addition Expressions

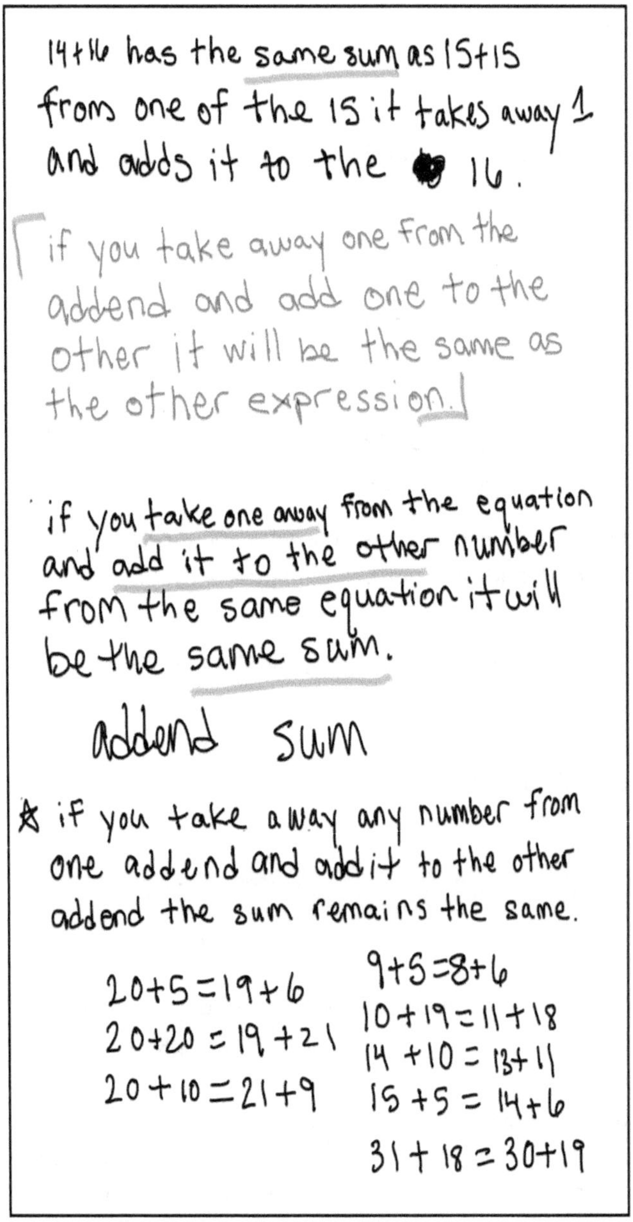

14+16 has the same sum as 15+15 from one of the 15 it takes away 1 and adds it to the 16.

if you take away one from the addend and add one to the other it will be the same as the other expression.

if you take one away from the equation and add it to the other number from the same equation it will be the same sum.

addend   sum

✷ if you take away any number from one addend and add it to the other addend the sum remains the same.

$$20+5=19+6$$
$$20+20=19+21$$
$$20+10=21+9$$

$$9+5=8+6$$
$$10+19=11+18$$
$$14+10=13+11$$
$$15+5=14+6$$
$$31+18=30+19$$

In a lesson before the one you'll see, the class had begun to think about equivalent expressions in subtraction. Ms. Clarke gave them some starting expressions, like 17 – 5, and asked them to figure out how to generate other subtraction expressions with the same difference. Students worked individually and in small groups on this task and came up with some sequences of equivalent expressions, like 17 – 5, 18 – 6, 19 – 7, or 17 – 5, 16 – 4, 15 – 3. For this next lesson, Ms. Clarke chose some of their sequences to discuss with the whole group. While the class had already had some experience with conjectures from their addition work, and some students were ready to move ahead to stating conjectures (as you'll see in the video), she felt many students would benefit from continuing to examine and describe patterns in the sequences of expressions. Given the complexity of the operation of subtraction, lingering on these patterns gives students time to discern the mathematical structure behind the patterns and to see that subtraction behaves in ways different from addition. We expect different students to be in different places in this complex territory. As you watch the video of the class's continued exploration, consider what different math ideas different students might be working on.

First Viewing of the Video:
"I Want to Build on My Classmate's Thinking"

Watch Video 5.1 with a focus on the mathematics students are learning.

Video 5.1

"I Want to Build on My Classmate's Thinking"

qrs.ly/exfs4xg

Reflecting on the Video:
The Mathematics Students Are Working On

[You may want to use the transcript at the end of this chapter as you consider these questions.]

1. Noticing patterns allows different students to work on different aspects of mathematics during the same discussion. List as many aspects of mathematics content as you can that different students might be working on. Then consider each of the students who speaks during the video clip: As a teacher, what would you want to remember about the math ideas each student offers?

2. Why do you think Ms. Clarke opens the discussion with the question, "What do they all equal?"

Second Viewing of the Video: "I Want to Build on My Classmate's Thinking"

Now rewatch the video clip with a focus on how students have voice during this discussion, how they build on each other's ideas, and what moves Ms. Clarke makes to support participation and collaboration.

Reflecting on the Video: Noticing and Collaboration

[You may want to use the transcript at the end of this chapter as you consider these questions.]

1. Identify places in the transcript where a student is connecting to another student's idea.

2. Identify places in the transcript where Ms. Clarke points out how students are collaborating.

3. In what ways does Ms. Clarke follow students' lead? In what ways does she guide the students' attention to important math ideas?

4. Guled and Emmaline begin to articulate what is going on more *generally*, rather than describing specific examples. It's likely that not all of the second graders are yet thinking about a general way to describe the patterns they are noticing. If that's the case, what do you think about Ms. Clarke's decision to repeat Emmaline's statement? What reasons might there be for her choice to linger on this statement?

Read and Reflect on What Others See in the Video

Let's return to the four aspects of noticing patterns that contribute to the principle, *collaboration supports student agency,* listed at the beginning of this chapter:

1. "What do you notice?" is an *invitation*, an opening for all students to enter into making sense of the underlying structures and behaviors of the operations.

2. As students notice and describe patterns, teachers are alerted to mathematical ideas that a range of students has to offer.

3. Noticing patterns in sets of related expressions leads naturally into developing conjectures about how an operation behaves.

4. Investigations into arithmetic structure, in which different students can work simultaneously on different aspects of the content, encourage collective and individual agency.

When our Critical Friends viewed the video "I Want to Build on My Classmate's Thinking," their comments centered around three ideas: noticing patterns as an entrance to important mathematics, noticing patterns as a gateway to participation and collaboration, and the nature of the moves the teacher makes as she responds to students' contributions.

1. Noticing Patterns as an Entrance to Important Mathematics

Yi Law Chan: The work that these kids are doing is not just about looking for patterns for the sake of looking for patterns, but there is something that's happening here that's going to lead them into thinking about something bigger about the structure of the operations.

Cindy Ballenger: It seems like the ideas get increasingly complex. The first kids all say this one's increasing by 1 and this one's increasing by 1. Somebody says they're *both* increasing by 1, and somebody else says *because* they're both increasing by 1, the difference stays the same. And Guled says, if they weren't both increasing by the same amount, it wouldn't stay the same. It's really building blocks of ideas.

Marta Garcia: I think this clip illustrates how a discussion on a set of expressions moves from just noticing the changes in the numbers to attempts to explain why those changes are happening. That is at the heart of the work of moving from noticing to conjecturing, and it illuminates how in this case subtraction is acting differently than addition. We can see how some students, like Cianna, are adding to what they are noticing even as they articulate the pattern. That contributes to beginning to see a general claim or structure.

Hetal Patel: What does it mean to do math? We often think it's about finding the answers, but, no, the *start* of this is that they all equal 12—what it means to do math is to unpack what's going on here. What's important is to notice and unpack patterns and relationships. That's how the important math ideas are elevated and grounded. That's what's important in terms of thinking about the question, What does it mean to do math?

2. Noticing Patterns as a Gateway to Participation and Collaboration

Yi Law Chan: Noticing patterns is an invitation. It's about access. Everyone can notice something, right? It's an invitation for all, just asking students to share what they noticed. We're able to hear a lot of different students' voices, but it's not as if these students' ideas are discreet and separate. The students are invited to share their noticings, but they're all also focused on growing an idea or ideas about whatever it is. What is the teacher's role in doing that or supporting that, and what are the students' responsibilities to attend to what others are saying, so that it *can* grow? It's not just *my* noticing separate from my classmates' noticing.

Virginia Bastable: I'm specifically thinking of this as collaboration. Each student has something to say, but it's different from what all the other students say. It feels like an example of the whole being bigger than the sum of the parts. One student talks about counting up and counting down. And somebody else talks about taking away 1 from both numbers, bringing in the operation of subtraction. Then somebody else says that, in the other set, both numbers are increasing by 1. It feels like that's an example of how all of the students' noticings together add up to something that can be more coherent.

3. The Teacher's Moves in Response to Each Student's Ideas

Lynne Godfrey: Noticing patterns is a gateway into the rigorous mathematics. That's what's happening in this class. And I was noting all of the teacher decisions or teacher moves during that time. I think I listed maybe 15 to 20 of them.

Hetal Patel: Each time she makes the decision to restate she checks in with the students, saying, "Hey, correct me if I'm wrong" or "Is this what you meant?" or "Does this represent your thinking?" And it appears that she does it in places where she wants to linger on the math idea for a little bit longer. I noted that right after Guled spoke, she said, "I heard two ideas in what Guled just said, so I want to repeat them so that you can think about it." She recognized that as a space to linger. She's letting the kids know, hey, this is something we can think about together. It's interesting to think about when teachers make the choice to have *other kids* restate their ideas or *the students themselves* restate their own ideas versus *the teacher* restating the ideas. When do you make which decision, and why?

Virginia Bastable: There's a student who indicates they couldn't take in what Emmaline says, and Ms. Clarke asks the class, "Do you want her to say it again?" That's a move that says, I'm expecting you to understand.

I thought it was also really worthwhile to highlight the fact that the teacher doesn't correct Guled when he uses the word *sum* incorrectly. She doesn't jump right in and say, oh, it's important to use correct math vocabulary. His idea is so perfect—you know what he's saying. But when she rephrases the idea, she uses the word *difference*. It's something that's worth highlighting because I think some people will go away thinking she shouldn't have let that go by.

Lynne Godfrey: Back to what Virginia said, I just noticed that Guled did say *sum* in the beginning of his statement, then he corrected it and said *difference*, but then he ends up saying *sum* again. So I think that's a clue that Ms. Clarke picks up on that he knows what he's talking about. I agree there's no need to correct him to use the right word in the moment because it was the *idea* that needed to be elevated for the class.

Ms. Clarke spoke to her own thinking about this in a piece of writing about this lesson:

Quayisha Clarke: When Guled explains what is the same, I think he might have our conjecture about addition in mind and, from habit, mentions the *sum* instead of the *difference*. His idea is correct, though. I know Guled as a mathematician and as a thinker, and I want him and all students to feel comfortable putting into words these big mathematical ideas and so I don't harp on his choice of the wrong term. Instead, I make the change when I restate his idea for all students to hold onto and hear a second time. I'm validating his thinking and shifting his language to what he thought he was saying and what he started to say. This is a minor teaching move with a major impact.

Students need to feel comfortable and able to take risks. They need to see their friends trying to use the mathematical language and stumbling a bit and then revising their thinking and moving on. I had a choice in how I addressed this, and I did so in a way that I thought would keep the momentum going with the mathematical contribution Guled was making to this discussion.

Reflection Questions

1. How is noticing and describing patterns a gateway for students' learning, participation, and collaboration?

2. Our commentators were interested in considering the many moves that Ms. Clarke makes in response to her students' ideas. Identify five moves that Ms. Clarke makes during these seven minutes and think about these questions for each of them: (a) Why might Ms. Clarke have chosen to do this? (b) What do you think the impact might be on individual students and/or on the class as a whole?

3. Review your responses to Question #2. Are there one or two of these moves that strike you as particularly important to reflect on in relation to your own practice?

What Do You Want to Remember From This Chapter?

Take a few minutes to note for yourself ideas you want to hold onto as you continue to investigate the meaning of a mathematics community and how to build it. What teacher moves have you noticed in this chapter that you want to bring into your own practice? What have you noticed about what supports a view of mathematics learning as collaborative? Here are a few aspects of discussions about noticing patterns that we see enacted in this chapter:

- **Open discussion with an invitation for multiple contributions.** The question, "What do you notice?," is open-ended and accessible. Because the answers to the problems chosen are not at issue, more students may feel comfortable offering their ideas.

- **Refrain from correcting every error**. When a student is working hard to get out a complex thought, that may not be the time to correct computation errors or incorrect vocabulary. Keep the focus on the student's main idea. Errors can be addressed at another time, by rephrasing, or by a quick check-in, especially when you're pretty sure the student has simply misspoken (e.g., "Did you mean this adds up to 12 or 13?"). Take into account what you know about the student, whether the error is just a slip of the tongue or something of mathematical significance, and what is needed to enable others to understand what the student is trying to express.

- **Check in frequently that each student's contribution is understood.** Ask classmates to paraphrase, clarify, or give an example. Ask the students who offered the ideas to confirm how their ideas are being recorded or whether another student's paraphrase captures what they meant. These moves both communicate to the individuals the value of their ideas and signal to the class the importance of listening with engagement and curiosity.

Taking a Next Step

Engage students in an activity about noticing patterns about an operation by listing a sequence of related expressions or equations and asking students what they notice. Start with addition and with computations that are easily solved by the students. For example, you could use examples of equivalent addition expressions (from Chapter 4):

SET A	SET B
$10 + 5$	$10 + 5$
$9 + 6$	$11 + 4$
$8 + 7$	$12 + 3$

Ask questions that emphasize listening to each other and extending the conversation: "Who can add on to what Alicia said?"; "Did everyone hear and understand what Edouard is saying? Who can say what you think Edouard is saying?"; "Who else noticed something about this set of expressions?" You might then choose a pattern with the same structure, but different numbers, or ask students to generate their own set of expressions that works the same way.

Video 5.1 Transcript: "I Want to Build on My Classmate's Thinking"

Ms. Clarke: We're returning back to our subtraction expressions. So if you take a look at all of these expressions, what do they all equal? What are these expressions equal to? Emmaline?

Emmaline: They're all equal to 12.

Ms. Clarke: They're all equal to 12. And I collected your subtraction expressions, and I was able to put here [points to chart] a sequence. So here we have one sequence of subtraction expressions that some of you came up with, and here we have another subtraction sequence of expressions that you all came up with. I want you to think for a minute, sitting still, what do you notice about these subtraction expressions. And you can look at them as this is one sequence [indicating left-hand column of expressions] and this is one sequence [indicating right-hand column of expressions]. So what do you notice? Give me a silent thumb on your chest. What do you notice?

[Pause]

Why don't we just popcorn out? What do you notice about these subtraction expressions? Aven?

Aven: I noticed that with this one [indicating left-hand column] it's counting down this way, 17, 16, 15, 14 and 5, 4, 3, 2. And in this one [indicating right-hand column] it's counting 17, 18, 19, 20 and 5, 6, 7, 8.

Ms. Clarke: OK. Does anyone want to add on to what Aven is saying? So here she said on this side it's counting down. That's what I heard you say. And on this side, you said it's counting up?

Aven: And it's kind of like, where you like compare [unintelligible]. Then they would all be equal—18 minus 6 is 12, 16 minus 4 is 12. But what I notice is that they all equal 12.

Ms. Clarke: Right, so we want to be very clear that all of these expressions equal 12. What else do you notice? Cianna?

Cianna: I noticed that each equation they take 1 from both numbers.

Ms. Clarke: Can you show me that?

Cianna: They're taking away, like the 17, from the 17 they're taking away 1 from the 17 and also 1 from the 5. That's how they got 16 minus 4.

Ms. Clarke: OK, so for this expression . . .

Cianna: Actually, all of them.

Ms. Clarke:	All of these?
Cianna:	Yeah, and then you take away [unintelligible]. [points to another expression on the chart]
Ms. Clarke [recording]:	OK, so we're taking away another one, and on this side as well. Is it OK if I mark it like this [**Cianna:** Yeah] to show where it's decreasing these by one?
Ms. Clarke:	OK, someone else. Emmaline?
Emmaline:	I noticed that on this one it's increasing.
Ms. Clarke:	OK, can you give me an example?
Emmaline:	The 17 is increasing by 1 to 18.
Ms. Clarke:	OK, so you said this increases by 1. Is that OK, if I put plus 1 here?
Emmaline:	Uh huh.
Ms. Clarke:	That captures your thinking? Does anyone else want to add on what they notice? Nasenguer?
Nasenguer:	What I noticed is that the 17 is increasing by 1 just like the 5 increased by 1, too.
Ms. Clarke:	So you added on to Emmaline. So the 17 increased to 18 and the 5 increased to 6. So then, what's happening with our difference? Or what is happening with the total amount that we have left? Angelica?
Angelica:	The total amount that we have left stays the same.
Ms. Clarke:	How do you know that?
Angelica:	Because for 17 − 5, that equals 12 but also 18 − 6 equals 12, because you're just taking 1 from the 5, you have 1 from the 5. From 17 you actually increase 1 to 18.
Ms. Clarke:	So, when we increase 1, why does our difference stay the same? Guled, you want to build on?
Guled:	I'd like to add on to Angelica because our sum, I mean our difference, stays the same, because if we only added or subtracted to one of the numbers in the equation, then there would be a different sum. But if we added to both the numbers in the equation, then it would still be the same.
Ms. Clarke:	I heard two ideas in what Guled just said. So I want to repeat them, so that you can think about it. So, he said, and correct me if I'm wrong, you said if we add 1 to one number, then the difference would be different. But if we add 1 to both numbers, I think you said the difference will stay the same.
Guled:	Yeah.
Ms. Clarke:	Okay, does anyone want to share, build on? Emmaline? We're going to think about it.

Emmaline: So, I wanna build on Guled, because when you add 1 to both of the numbers, since it's subtraction, you're also adding 1 to the number that's subtracting. So then you're taking away the 1 you added to the other number.

Student: I didn't hear what she just said.

Ms. Clarke: Who thinks they can repeat what Emmaline just said? Do you want her to say it again? [Some students: Yes.] Emmaline, why don't you say that again and let's all focus in on her thinking.

Emmaline: I wanna build on what Guled, because when you add 1 to both of the numbers, since it's subtraction, the 1 that you're subtracting adds 1, and it's gonna take away the 1 you already added.

Ms. Clarke: Can someone say what she just said in their own words?

Student: I didn't hear.

Ms. Clarke: She said if you're adding 1 to the first number and you're adding 1 to the second number, you're going to subtract the 1 that you just added.

Student: Oh, yeah.

Ms. Clarke: That's what you said, right, Emmaline?

Chapter 6

Collaboratively Building Toward a Conjecture

Early in their study of multiplication, a class of third graders noticed a pattern: $4 \times 5 = 20$ and $5 \times 4 = 20$; $3 \times 2 = 6$ and also $2 \times 3 = 6$. When students started to describe the pattern as "switcheroos"—the same term they used when they had noticed a similar pattern when adding— the teacher suggested they try to come up with a sentence that described what they noticed to someone new to the discussion. "What if the principal came into our room? Would Mr. Martin understand what we mean?" It seemed that everyone in the class could see the pattern, but it was challenging to find the words.

One student offered, "If you write the problem backwards, the answer stays the same," but other students weren't sure what "backwards" would mean to someone who hadn't been part of their discussions. A second student said, "What if Mr. Martin thought we mean changing $3 \times 2 = 6$ to $6 \times 2 = 3$? That's not right." A third student proposed, "If you switch the numbers around, the answer stays the same." A fourth then wondered if it would be clear which numbers would be switched.

These third graders are beginning to articulate a *conjecture*. A conjecture in mathematics is a statement about something general that seems to have a good chance of being true for some set of mathematical objects—for example, all whole numbers. It is often based on studying examples of the idea and noticing the regularities across those examples, as the third graders had been doing, but it has not yet been proven as true for all possible examples.

We've found consistently that elementary students are fascinated with creating a conjecture that is clear and precise, that says what they want to say about the mathematics. This articulation is difficult and requires collaboration as statements are tried out and revised. It offers the opportunity for students to think about precision in everyday language and to learn mathematical terms that are useful to them in saying what they want to say.

The phases of noticing patterns and articulating conjectures based on those patterns often blend into each other, with different students working on different aspects of the mathematics. While some students are describing numerical patterns, other students might be thinking about a conjecture about that pattern. When Ms. Clarke lingered on Guled's and Emmaline's general claims in Chapter 5, she was both acknowledging that they had something important to offer and putting those ideas out to the rest of the class to consider as their study of subtraction continued.

The move from noticing to conjecturing also brings up another significant aspect of understanding the operations: that different operations behave differently. Ms. Clarke's second graders had initially been surprised to discover that equivalent expressions in subtraction had a different structure than the equivalent expressions in addition they had previously considered. Similarly, in this chapter, third graders find that another aspect of subtraction is quite different from what they had discovered about addition.

> **In this chapter, you will**
>
> - do some mathematics to explore a structure of subtraction and to prepare for understanding the work of the students in the video,
> - watch a video clip of third graders investigating patterns and developing conjectures in subtraction and consider what they are learning and how they are participating, and
> - read commentary from some of our Critical Friends and from the teacher about the video clip.

As you do these activities, we'd like you to think about these two aspects of how noticing and conjecturing contribute to the principle that *collaboration supports student agency.*

1. Noticing patterns and developing conjectures based on those patterns is a context in which teachers can encourage students to listen to each other and build on each other's ideas.

2. Articulating conjectures helps students to understand the operations as more than a set of procedures to remember and to come to see that each operation has a unique set of generalizations associated with it.

Do the Math

Here's a dilemma that came up in a fourth-grade class as the students considered how to solve 145 – 98 mentally. The students knew that 145 – 100 = 45, and they knew that there should be something about a change of 2 to get the answer to the original problem, but was the difference of 145 – 98 *2 more* or *2 less* than 45?

Subtraction is mathematically complex—for our students and even for us as adults. While those of you reading this have your own ways to solve 145 – 98, it wouldn't be surprising if you had a bit of hesitation in comparing 145 – 98 to 145 – 100. The two expressions are an example of something important about the operation of subtraction. Noticing and describing this pattern, and thinking about why it holds, is a more complex mathematical task than simply finding the answer to 145 – 98.

The expectation that subtraction or multiplication will work like addition can be the source of many calculation errors and, in later years, interfere with the learning of algebra, especially if students only memorize procedures and don't delve into the structure and behavior of each operation during their elementary years. You noticed contrasts between addition and subtraction when you worked on the math in Chapters 4 and 5.

In the operation of addition, you might think about two different kinds of values—the addends and the sum. But in subtraction, there are three different kinds of values to consider, each of which functions differently in the equation. Third graders, who often have an image of subtraction as removal, or "take-away," might think of these different values as the amount you start with, the amount you take away, and the amount left.

In order to investigate how addition and subtraction behave differently, let's start by looking at some sets of related addition problems.

SET A	SET B
3 + 5 = 8	3 + 5 = 8
3 + 6 = 9	4 + 5 = 9
3 + 7 = 10	5 + 5 = 10
3 + 8 = 11	6 + 5 = 11

1. What do you notice about Set A?
2. What do you notice about Set B?

(Continued)

(Continued)

3. Write out in words one or more statements about what is going on in Set A, in Set B, and/or in both. You might start your sentence with "If . . ." or "When. . . ."

If you are working with others, share your statements with your colleagues.

Now look at each set of four equations below:

SET A	SET B
$8 - 3 = 5$	$8 - 3 = 5$
$9 - 3 = 6$	$8 - 4 = 4$
$10 - 3 = 7$	$8 - 5 = 3$
$11 - 3 = 8$	$8 - 6 = 2$

1. What do you notice about Set A?

2. What do you notice about Set B?

3. Write out in words one or more statements about what you think is going on in Set A, in Set B, and/or in both. You might start your sentences with "If . . ." or "When. . . ."

If you are working with others, share your statements with your colleagues.

4. Compare the statements you wrote about the addition sequences to the statements you wrote about the subtraction sequences. What do you notice about how they are similar and how they are different? Discuss this with your colleagues.

Watch the Video: "What Happens to the Difference?"

You are going to watch Emmanuel Fairley-Pittman's third-grade class working on changing a number in a subtraction expression. In prior lessons, students had worked on adding 1 to an addend and eventually had come up with a class conjecture: *If you add 1 to an addend and keep the other addend the same, the sum will increase by 1.* When they began to examine analogous patterns in subtraction, they saw that something different was going on. In a lesson shortly before the one you'll see in the video, the students had considered the effect on the difference of increasing the first number in a subtraction expression—the minuend or the amount you start with—as illustrated by the two story problems in Figure 6.1.

Figure 6.1 • Two Story Problems That Illustrate the Effect on the Difference of Increasing the Minuend (the Starting Amount) by 1

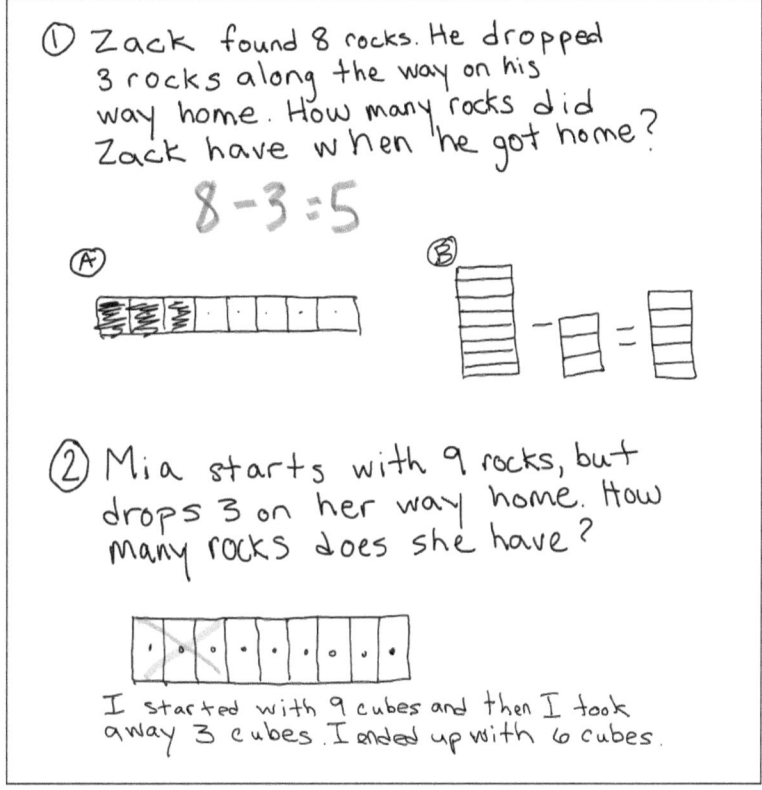

The small numbers were deliberately chosen so the calculations would be well known to the students. While this work on noticing and describing patterns invites students in, it is also challenging—not a time to be using numbers for which the computation itself is difficult. After working with this and similar problem pairs, the third graders came up with a conjecture (see Figure 6.2).

Figure 6.2 • The Students' Conjecture About Increasing the Starting Amount by 1

In a subtraction problem,
minuend
if (the amount I start with) increases by 1

and (the amount I subtract) stays the same
subtrahend
then the difference increases by 1.

The teacher had introduced technical mathematical vocabulary as students wanted and needed it, and this less familiar language was also noted on the poster of their conjecture. Both the math vocabulary and the everyday vocabulary were retained together, so that students could choose the words they were ready to understand and use.

We're going to join the class during a later lesson as they're working on what happens to the difference if you increase the second number, the *subtrahend*, or *the amount taken away*. In this lesson, students have been discussing the example 8 – 3 and 8 – 4. Mr. Fairley-Pittman felt that students needed to ground their words in a representation and also that they needed to get up from the rug and move. He sent them off in pairs and asked them to use cubes to represent 8 – 3 and 8 – 4, not just to show what's going on in this particular pair of problems, but to think about what this is an example of, and to come up with a general statement, a conjecture, about subtraction. Note that grounding students' thinking in pictures, diagrams, and models is another important facet of building participation and collaboration in mathematics class; we'll delve deeply into the importance of representation in Part Three.

The video clip begins with the last 25 seconds of students working in pairs. While you won't be able to hear what students are saying, pay attention to what is going on in the classroom—students' actions, gestures, and expressions. Then the class comes together to share their thinking. Note that this is the very end of the school day, and you'll hear an announcement asking for one of the students to be dismissed.

First Viewing of the Video: The Mathematics Students Are Working On

Watch the 4-minute video clip, Video 6.1, "What Happens to the Difference?," with a focus on what students are learning.

Video 6.1

"What Happens to the Difference?"

qrs.ly/wtfs4y4

Reflecting on the Video:
The Mathematics Students Are Working On

[You may want to use the transcript at the end of the chapter as you consider these questions.]

1. What key math ideas for third graders are students working on?

2. Why do you think these third graders are animated and engaged as they discuss simple subtraction, like 8 − 3 and 8 − 4, when they already know the answers?

3. Do you see evidence that students are beginning to see 8 − 3 and 8 − 4 as one example of something more general about subtraction? Can you identify this in students' language?

Second Viewing of the Video: Supporting Individual and Collective Agency Through Noticing Patterns

Re-watch the video clip "What Happens to the Difference?" with a focus on how students have voice during this discussion.

Reflecting on the Video:
Students' Individual and Collective Agency

[You may want to use the transcript at the end of the chapter as you consider these questions.]

1. How does this discussion support students' collaboration?

2. Are there particular moves the teacher makes that promote students' identity and agency in mathematics?

Read and Reflect on
What Others See in the Video

Let's return to the two aspects of how noticing patterns and developing conjectures contribute to the principle that *collaboration supports student agency* listed at the beginning of this chapter:

> 1. Noticing patterns and developing conjectures based on those patterns is a context in which teachers can encourage students to listen to each other and build on each other's ideas.
>
> 2. Articulating conjectures helps students to understand the operations as more than a set of procedures to remember and to come to see that each operation has a unique set of generalizations associated with it.

In this section, Critical Friends comment on developing conjectures as a context for collaboration and on the teacher's role in establishing a collaborative community. Mr. Fairley-Pittman reflects on the importance of student collaboration in his practice, and about how listening and collaboration in math class is tied to his beliefs about equity and access for his students.

1. Formulating Conjectures as an Invitation to Collaborate

Darlene Ratliff: This is what a math community looks like— kids talking to each other, kids building on ideas with each other. The teacher is facilitating the conversation, not telling them anything. They're building on each other's ideas and developing their own conjectures.

Virginia Bastable: I noticed that we don't see the teacher rephrasing all the kids' conjectures. The kids' conjectures stand on their own and they listen to each other. They don't need him to say it again for them. They're not waiting to hear what the teacher has to say.

Near the end of the video, Mr. Fairley-Pittman calls on Hailey at the end of class, even though she has been called by the front office to be dismissed. Critical Friends paid particular attention to this move and how Mr. Fairley-Pittman responded to Hailey's contribution. Were his actions consistent with a goal of student collaboration?

Lynne Godfrey: How well you have to know your students, and how well he must have listened to them as they were thinking through these ideas, that he could call on Hailey in that moment and knew she'd have something to say. And she does. She says, "My thought is that . . ." And that's not an easy move to make. He knew something about her, knew that he could call on her in that moment.

Marta Garcia: What's important to me about the teacher's enthusiastic reaction to Hailey is the fact that he then asks Kedly to weigh in. The teacher himself doesn't weigh in; another student weighs in. That's the distinction that made it OK for me. It felt to me like the excitement was that the children are building on each other's ideas. He's not saying, OK, Hailey said this, let's write it on the board. He goes back to Kedly. Hailey's idea is born of Rollie's and Johansen's work. It's not just her idea.

We're celebrating a collaboration of ideas. In this video, I see an example of a classroom community that values a collaborative building of ideas, one that includes multiple forms of language (such as everyday language, gestures, the use of representations). In spaces such as this, multilingualism can be invited and seen as a strength. While we didn't see specific students using languages other than English, we can imagine the opportunities there would be for multilingual students to engage in deep and rigorous mathematics as they use their linguistic resources to express their ideas.

2. Persistence in Developing Collaboration

Emmanuel Fairley-Pittman: I tell students all the time that I am not the only teacher in the room and that they all can learn something from each other when they listen. With a classroom of 25 students, it's imperative that students know and use each other as resources for learning. I, their teacher, am not always available to address their questions and might also misinterpret their questions. When they talk and listen to one another, they get another perspective that might aid in their learning in ways that I cannot or am not available to do.

Moreover, I have learned that the work of having students talk to and listen to one another is something that develops over the course of the school year and is never "mastered" by any student. To be quite frank, I don't know what "mastering" this skill looks like. Even I, a 30-something-year-old, am still learning and defining for myself what it means to talk and listen productively. That said, it's important that students work on these skills because they will use them at every level of their learning in every subject. It's even more important that they practice these skills in math because math is such a complex content area. The more advanced the math becomes, the more they can, and hopefully will, lean on the understanding of others to make sense of their own mathematical understanding.

3. Pointing Out Acts of Collective Agency

Yi Law Chan: I noticed Mr. Fairley-Pittman doing something at the end that I think is really valuable in classrooms—helping kids to recognize their work as mathematicians. He said to Hailey, "Correct me if I am wrong, Hailey, were you listening to Rollie and Johansen to help you come up with this idea, this conjecture?" It's an affirmation of what the student is doing, how it's helping all of us to learn, and articulating what the work of a true mathematician is like.

Hetal Patel: The ending stood out to me, too, when he was naming the attributes of what it means to engage in math—using vocabulary, listening, using the tools in the room, and those tools include your classmates. Your classmates' comments are part of the tools that you're connecting to. Your job as a learner and as a community member is to listen and connect.

4. Collaboration on Challenging Mathematics to Support Equity and Access

Emmanuel Fairley-Pittman: To anyone who believes students like mine—students of color—are not capable of engaging with rigorous math concepts, I would say that it's not the students, it's the teaching. I would argue that for the people who believe rigorous mathematics is only for certain students, or a certain type of student, they have not taken the time or made the effort to modify their practice so that all students can engage with this type of work.

Equitable, rigorous mathematics teaching should be exactly what it claims to be: teaching that supports all students. It's my job to make sure students are doing the majority of the mental "heavy lifting" when we are in math class, but that also means I have to do some mental "heavy lifting" before and after to ensure that all of my students are able to engage. I can't just show up to a lesson and expect that students are going to just know what to do.

Orienting students to each other's thinking and facilitating discussion between students about complex mathematical concepts is a necessary and equitable practice because it provides students with access points to engage with the ideas being presented. When students have a chance to talk with their peers and are encouraged to listen to them, an environment of mutual respect, accountability, and challenge is formed.

> Because all students are new to the ideas, there is a level of freedom that students feel, which they don't feel when they are simply having conversations with adults who may already understand the concept and who are likely to have a specific way of approaching a problem. Students are able to make sense of the ideas, making errors, connections, and adjustments to their thinking, while being guided by me, their teacher. In this process, they begin to embrace their identity as a mathematician. That identity is bigger than any one mathematical concept and will follow them beyond my classroom.

Reflection Questions

1. What are aspects of this lesson—the content of developing a conjecture, what the teacher does, and what the students do—that support equitable participation and students' access to making sense of important mathematical ideas?

2. In Chapter 2, we talked about the importance of teachers communicating curiosity about students' mathematical ideas. Once we start truly listening for the kernels of sense-making in students' ideas, we are often excited by what they are noticing and articulating. What are the pros and cons of a teacher reacting with enthusiasm to a student's ideas? What would you want to consider in monitoring your own excitement about student ideas?

3. Look through the transcript, focusing on what the teacher says. What do you think he is trying to accomplish with his opening remarks, his questions, and his summary remarks at the end of the lesson?

What Do You Want to Remember From This Chapter?

The mathematics of noticing patterns and developing conjectures offers both important mathematics content and a context that invites student participation. Before leaving this chapter, take a few minutes to jot down one or two ideas you want to take away as you think about your own teaching or the teachers you support. We offer a few thoughts that have been important to us as we considered the work of the third graders in this chapter and the comments of their teacher and of our Critical Friends:

- **Practice collaborative discussions**. Learning to have discussions in which students listen to and build on each other's ideas requires practice and reminders, just like any other classroom structure.

- **Explicitly name for students how they've collaborated**. Characterizing what students are doing explicitly, as the teacher does at the end of the lesson in this chapter, reinforces for students what it means to enact collective mathematical agency.

Taking a Next Step

1. Mr. Fairley-Pittman challenges us to think about what it means to teach equitable, rigorous mathematics. Choose a sentence or two from his writing that challenges your own practice. In what ways do you want to reflect further on those sentences?

2. Revisit with your students what they noticed about the equivalent addition expressions in "Taking a Next Step" in Chapter 5. Ask them to formulate a conjecture, that is, to come up with a sentence that describes the pattern they noticed. You might want to have students work individually or in pairs first and then come together to share some ideas.

Video 6.1 Transcript:
"What Happens to the Difference?"

[The video begins with 25 seconds of students working in pairs.]

Mr. Fairley-Pittman:	Alright, so what happens to the difference? Let's see if we can come up with a conjecture. What happens to the difference when the subtrahend changes? Can we all listen to people? Don't talk at me, talk to each other, please. Johansen, you start the conversation but talk to your classmates.
Johansen:	The difference gets smaller, because you just added one more to the 3 to make it a 4 minus 4 equals 4. Rollie?
Mr. Fairley-Pittman:	You've got to speak really loud, Rollie.
Rollie:	I agree with you that the difference gets smaller when the subtrahend gets bigger because that you were taking away 4, not 4, you were taking away more. 'Cause in 8 minus 3 you were only taking away 3. So for 8 minus 4, it's going to be minus, the difference is gonna be 1 less because you're taking away 1 more because the subtrahend is 1 more, so the difference has to be 1 less.
Mr. Fairley-Pittman:	Before you go, do you have thoughts on this? Can you share with the class really loud?
Hailey:	My thought is that when the minuend stays the same and the subtrahend gets bigger, the difference gets smaller.
Mr. Fairley-Pittman:	Say it again?
Hailey:	I said when the minuend stays the same and the subtrahend gets bigger, the difference gets smaller.
Mr. Fairley-Pittman:	Kedly, you're agreeing with her. What did she say?
Kedly:	Hailey said when the minuend stays the same and the subtrahend changes, and the subtrahend gets bigger, the difference gets smaller.

Mr. Fairley-Pittman: Guys, you just came up with your own conjecture. When the minuend stays the same and the subtrahend gets bigger, the difference gets smaller. We are going to pause right here, what I love about, and we are going to come back to this, don't worry. Look, look, look, look, look. What I love about what Hailey just said is that she not only used the math language, she referenced what we were talking about earlier at the start of the lesson. And I do believe, correct me if I am wrong, Hailey, were you listening to Rollie and Johansen to help you come up with this idea, this conjecture? So, she was also listening to her classmates and thinking about what Rollie and Johansen said, using vocabulary, using the tools in the room to help her think about it. That is the work of a true mathematician, and I cannot thank you enough.

Chapter 7

Centering Students' Ideas to Formulate a Class Conjecture

In the previous chapter, as Mr. Fairley-Pittman's students discussed the pattern they noticed when comparing pairs of subtraction equations, they began to formulate conjectures. Rollie said, "The difference gets smaller when the subtrahend gets bigger. . . . The difference is gonna be 1 less because you're taking away 1 more," and Hailey offered, "When the minuend stays the same and the subtrahend gets bigger, the difference gets smaller."

Throughout this book, we emphasize the need for time, patience, and support for students to formulate and express their ideas. We've seen that finding the words to express a mathematical idea with clarity is challenging. In this chapter, we take a look at collaborative work centered on comparing and refining individuals' statements to come up with a "class conjecture." Our collaborating teachers realized the value of having students come up with a class conjecture, and they saw that it took the whole class engaging in hard work to get their statement right. They discovered that students become invested in this process and are willing to stick with crafting the precision and accuracy of their conjecture until they are sure it says what they want it to say. First, students need to have the experience and time, perhaps individually, perhaps in pairs, of coming up with their own conjectures. The teacher can then select several statements to present to the class, and drawing from those statements, students work together to create a class conjecture. During this process, students challenge each other to clarify ambiguities and revise wording. The teacher may also ask about the meaning of their words, raise questions about whether they've included everything that is relevant, or suggest technical vocabulary as the need arises. For example, if students use the word *number* to refer to different objects in their claim, the teacher might ask students how someone reading their words will know which number they're referring to.

In this chapter, which focuses on the activity of formulating and refining conjectures, you will

- write your own conjecture about changing a factor in a multiplication expression,
- watch a video of third graders discussing draft conjectures written by five different students,
- read the teacher's description of his goals for three students and see how those goals play out in the classroom, and
- consider commentary by our Critical Friends.

As you consider the classroom video and the commentaries in this chapter, we'd like you to think about these aspects of the principle that *collaboration supports student agency*:

1. Students' individual work on developing conjectures arises from noticing patterns. This individual work becomes a tool for moving the collective thinking forward.

2. In the service of producing a class conjecture, the teacher presents to the class a variety of statements from students' written work, not just those that are already most general and well-articulated.

3. As students ask questions about each other's conjectures, they collaboratively engage in developing mathematical statements that communicate with clarity and precision.

Do the Math

Consider these pairs of related equations:

$7 \times 3 = 21$ $7 \times 3 = 21$

$8 \times 3 = 24$ $7 \times 4 = 28$

1. Come up with several more sets of multiplication equations that are related in the same way.

2. Complete the statements below:

When I add 1 to the first factor, then this happens to the product: _____.

When I add 1 to the second factor, then this happens to the product: _____.

Share your statements with a colleague. What is the same about the statements you and your colleague created, and what is different?

Watch the Video: "She Wrote What It Meant"

In Mr. Fairley-Pittman's third-grade classroom, students have been investigating how adding 1 to a factor affects the product in a multiplication equation. Prior to this class, students worked independently to write their conjectures, referring to examples they had considered as a class, including 7×3 and 8×3. Mr. Fairley-Pittman selected the following five students' statements for the class to consider in this session.

Jaslanie: When I add 1 to the first factor, the product gets bigger because we add one from the 7 to the 8.

Kevin: When I add 1 to the factor, the product increases by 3.

Kedly: When I add 1 to the first factor, the product increases by how much we're multiplying by.

Rollie: When I add 1 to the first factor, the product increases by the second.

Hailey: When I add 1 to the first factor, the product increases by one of the second factor.

Mr. Fairley-Pittman presents the five statements—one by one—inviting students to discuss and pose questions about each. The class first reads and comments on Jaslanie's and Kevin's statements. We join the class as the teacher asks a student to read Kedly's statement.

First Viewing of the Video: "She Wrote What It Meant"

Watch Video 7.1, "She Wrote What It Meant," with a focus on what students are learning.

Video 7.1

"She Wrote What It Meant"

qrs.ly/9gfs4y7

Reflecting on the Video: What Students Are Learning

[You may want to use the transcript at the end of the chapter as you consider these questions.]

1. What does each of the five student statements convey about multiplication?
2. In what ways are the five statements the same, and in what ways are they different?
3. What might each statement contribute to the understanding of the class?
4. In Mr. Fairley-Pittman's comments at the end of the lesson, he suggested students consider whether the words they're using in their conjectures are "as precise as possible, while also being kind of general." What does that mean to you?

The Teacher's Profiles of Three Students

Before viewing the video the second time, read Mr. Fairley-Pittman's profiles of three of his students: Kedly, Hailey, and Jaslanie. He describes the different goals he sets for each individual, drawing on what he understands about each of their strengths and where they have room to grow. Then, in your second viewing of the video, look for these three students' participation and think about how Mr. Fairley-Pittman may be supporting the goals he has for each of them. (Note that Mr. Fairley-Pittman wrote these profiles the following year when he had many of the same students in Grade 4.)

Emmanuel Fairley-Pittman: I had the privilege of working with the same group of students for two years (Grades 3 and 4) and have been fortunate to watch them grow in many ways as mathematicians. I have selected three students to describe how they engaged in the lesson sequences on generalizing about the operations in markedly different ways: Kedly, an eager, astute student who enjoyed the challenge in every session to make sense of problems via representations and share his thinking; Hailey, an enthusiastic, dynamic and persistent student who took advantage of every discussion and opportunity to share; and Jaslanie, a more reserved, cautious student who grew in confidence and ability as we progressed through each session over the course of the two years I was her teacher.

Kedly. The lesson sequences on generalizing about the arithmetic operations were a chance for Kedly to learn and practice what it means to listen as a mathematician. They were also time for him to develop his critical reasoning and explanation skills. A generally outspoken student, these sessions challenged what Kedly thought he knew about mathematics and really forced him to wrestle with developing mathematical arguments and take into account others' critiques of his reasoning.

As Kedly moved through this work with his classmates, I observed him lean into the responsibility of mathematicians to not only know but also explain and grow ideas. Over the course of the two years, Kedly began to appreciate using visual representations as a resource to explain his thinking. He also began to see his classmates as more of a resource, as he participated in the collective math reasoning apparent during class discussions. I believe he understands more fully that mathematical thinking does not have to be a solitary pursuit but can be conversational and that dialogue is as important as the answer.

Hailey. Hailey is another student who really thrived during the lesson sequences about the operations. As one who typically enjoys a challenge, she used these sessions to wrestle with complex mathematical ideas. Over the course of our two years together, Hailey took advantage of the opportunity to sit with, discuss, and share her ideas about the different concepts we debated during our sessions.

My hopes for Hailey as we moved through each session were that she would continue to share what she was thinking and be willing and confident to engage with her classmates when things did not make sense to her. I wanted her, like all of her peers, to know that girls, specifically girls of color, not only "do math," but excel in math. I wanted her to set an example for her peers in the realm of mathematics, as she does in other content and social settings. Hailey did not disappoint. She would often take time to make colorful and detailed representations that she would use to explain her thinking. You could also see in each session her growing confidence as a mathematician. Whether it was sharing her own thinking or debating the thinking of her peers, Hailey seized every opportunity to respectfully make her voice heard in our discussions.

Jaslanie. Jaslanie was a more cautious student whom I have seen blossom over the past two years into a confident mathematician, willing to take risks during the lesson sequences on generalizations about the arithmetic operations. Over the course of the two years, I was very intentional about finding opportunities to share Jaslanie's thinking with her classmates, thereby providing a platform for her to grow her thinking.

Take, for example, one of our final sessions from Grade 3 in which we were discussing what happens when you increase one of the factors in a multiplication problem and keep one the same. The conversation began by posting individual student conjectures and progressively discussing the more complex ideas of each conjecture. Jaslanie's conjecture was the first idea that we unpacked as a class: *"When you add 1 to the first factor, the product gets bigger because we add 1 from the 7 to the 8."* As the conversation continued, the student conjectures became more complex and precise in language, and I observed Jaslanie not only participating but also synthesizing a critical idea expressed by her classmates. Toward the end of the discussion, Jaslanie explains, *"Rollie wrote 'the second,' but then Hailey, she wrote like what it meant. . . . So Hailey wrote what 'second' means."*

By using Jaslanie's thinking as the initial point of conversation for our class discussion, I believe Jaslanie saw her thinking as validated and felt more confident to speak up. Over time, Jaslanie became more engaged and not only shared her thinking but also synthesized the ideas of her peers and asked questions to better make sense of concepts and ideas we were discussing in class.

Second Viewing of the Video: "She Wrote What It Meant"

Watch the video clip, "She Wrote What It Meant," with a focus on how students have voice during this discussion. Notice how Kedly, Jaslanie, and Hailey participate.

Reflecting on the Video: How Students Have Voice

[You may want to use the transcript at the end of the chapter as you consider these questions.]

1. How do you see Mr. Fairley-Pittman's goals for Kedly, Hailey, and Jaslanie enacted in the lesson?

2. What are your observations or questions about how Mr. Fairley-Pittman maintains both goals for individual learners and a sense of mathematics community for the whole class? What are questions you want to think about in your own practice with regard to balancing support for individual students and maintaining a collaborative community?

Read and Reflect on
What Others See in the Video

Let's return to the aspects of how developing conjectures contribute to the principle that *collaboration supports student agency* listed at the beginning of this chapter:

---

1. Students' individual work on developing conjectures arises from noticing patterns. This individual work becomes a tool for moving the collective thinking forward.

2. In the service of producing a class conjecture, the teacher presents to the class a variety of statements from students' written work, not just those that are already most general and well-articulated.

3. As students ask questions about each other's conjectures, they collaboratively engage in developing mathematical statements that communicate with clarity and precision.

When our Critical Friends viewed the video "She Wrote What It Meant," their comments focused on three main themes:

1. The nature of the mathematics the students were doing

2. The different goals Mr. Fairley-Pittman set for different students and how those were integrated into the development of a collaborative mathematics community

3. His goals for Hailey's identity and agency as a Black girl engaging in mathematics

1. The Nature of the Mathematics

On the surface, it might seem that Mr. Fairley-Pittman's class was working on typical third-grade content: understanding basic multiplication. However, the class's task of articulating a conjecture highlighted for our Critical Friends that these students were engaged in mathematical concepts and practices that go beyond most K-5 programs. Specifically, they were working toward the distributive property of multiplication over addition, a property that is foundational to understanding and calculating with multiplication.

Lynne Godfrey: This content is usually reserved for middle-school students, but these third graders are capable of this kind of thinking. It ups the expectations that teachers can have. For now, they find their own language to describe what's happening. Maybe it's not formal mathematical language, but it's the language available to them. As they work at it, their language can get more precise.

Marta Garcia: If someone is looking at this video without having done any work with generalization or students making conjectures, they might wonder about how much time is spent on 8 × 3. But it's not 8 × 3 they're working on. They're working on a bigger idea. We can see the impact of these kinds of discussions on students' ability to question each other, and to think about justification—explaining *why* their conjectures are true. But it's also about computational fluency. If they understand how multiplication works, and if they don't remember one fact or a step in a calculation, they can figure it out.

Virginia Bastable: Some people refer to this content as "early algebra." But it's not just about making sure that all these kids do well when they get to ninth-grade algebra class. It's about helping them understand the math of their own curriculum more deeply so that they're more able to use it flexibly and comfortably.

Yi Law Chan: It was interesting to see the choices the teacher made of the different statements that were shared and how they were revealed one at a time. What the students understood about the mathematics unfolded throughout the discussion, with the students asking questions of one another and the students responding to one another. I was curious if students saw their own thinking in each other's work. What did they think about how the conversation evolved? Did it grow their own understanding of what happens when you change a factor?

2. Different Goals for Different Students

Mr. Fairley-Pittman wrote profiles for only three of his students, but he had goals for each member of the class based on their particular strengths and needs. After reading Mr. Fairley-Pittman's profiles, our Critical Friends were struck by how, within the context of a single discussion in which the teacher orchestrated movement toward a set of mathematical principles, he was also managing different goals for individual students. They considered implications for Jaslanie, Kedly, and Hailey and also addressed what Mr. Fairley-Pittman's goals meant for the class.

Hetal Patel: When I first viewed the video, I was worried that always going in order from the most accessible statement to the most general, over time, the class may know that, hey, if I'm first, mine is the most accessible, and if I'm last, that's the conjecture and generalization with the best wording. So how does that impact participation? There's a possible stigma for going first. But when I go back to what the teacher's goals were for Jaslanie and I watch the video again, I see that Jaslanie stayed engaged throughout the whole discussion. She was asking questions, and she was the one who clarified the difference between Rollie's and Hailey's statements. I realize all of the statements contribute to the collective conversation. Even though they may have varied levels of specificity, they all contribute to the community goal in a valuable way, and Jaslanie knew that.

Virginia Bastable: I was noticing the way different students' questions were placing demands on Kedly to explain more about what he meant. I was really struck by the fact that he was the one who was invited to go up to the board and point, because people were asking questions, and he wasn't sure how to answer them from his seat. So it felt like there was some pressure on him to explain his thinking to others and not just for himself.

The way Mr. Fairley-Pittman phrased his goals for these individual students, it was in terms of making the classroom a community of mathematicians. They each participate according to who they actually are. He's honoring each student and thinking about, given who they are and what he's seen them do already, how they can expand and deepen what they have to offer to themselves and to the class.

Darlene Ratliff: I was thinking about how much the teacher is holding onto. There's the math, and then there's everything he knows about his students. He's working to facilitate each person's learning at the same time that he's moving the mathematical ideas forward. And he's working to build the math community. There's a lot going on that is happening simultaneously.

3. Highlighting the Contributions of Black Girls

In the beginning of the year, Mr. Fairley-Pittman noticed strong voices among the boys in his class. He felt the need to attend to nurturing and highlighting the voices of girls, including Black girls—one example of a group that has been historically neglected in school mathematics.

Lynne Godfrey: I appreciated what Mr. Fairley-Pittman said about Hailey. If you're paying attention to the Black girl in your class, you're probably doing a better job of paying attention to everybody in the class. I've gone into too many classes where the Black girls were ignored as long as they were quiet and not disrupting.

Hetal Patel: That's a value that the teacher and the students take on, being mindful to elevate a range of students, including the Black girls, and it's going to generate learning for all.

Marta Garcia: Mr. Fairley-Pittman says that he wanted Hailey to set an example for her peers as a mathematician as she does in other content and social settings. If Hailey has high social status, then by modeling her interest in mathematics, she also opens that up for others.

Reflection Questions

1. Critical Friends noticed that while Mr. Fairley-Pittman's students were engaged in "early algebra," the ideas they were working on are essential to understanding third-grade content. How might formulating the class conjecture deepen students' understanding of multiplication, and how might it strengthen students' facility in calculation?

2. Virginia Bastable says, "The way Mr. Fairley-Pittman phrased his goals for these individual students, it was in terms of making the classroom a community of mathematicians." And Lynne Godfrey comments, "If you're paying attention to the Black girl in your class, you're probably doing a better job of paying attention to everybody in the class." In what ways can supporting individual students to develop agency as doers of mathematics contribute to the collective mathematical agency of the class?

What Do You Want to Remember From This Chapter?

The mathematics of noticing patterns and developing a class conjecture offers an important context for collaborative thinking. Take a few minutes to jot down one or two ideas based on the video and commentary in this chapter that you want to continue to think about. Here are three thoughts that have been important to us as we considered the work of the third graders in this chapter and the comments of their teacher and of our Critical Friends:

- **Choose students' work to share with intention.** Consider both the needs of individual students and the flow of ideas for the class. Presenting a piece of student work for the class to consider can encourage students to more actively share their thinking in class discussion. Often, beginning with a version of a conjecture that references specific numbers is a gateway to bring more students into the discussion, and they can continue to participate as the statements become more general.

- **Encourage students to question each other's ideas.** Participating in a collaborative mathematics community includes asking questions about or explaining a classmate's thinking.

- **Pay attention over time.** Develop ways to track which students are participating and in what ways. Are there factors, such as gender, ethnicity, race, neurodiversity, or language that are affecting whose voices are heard? Who speaks up in whole-group discussions? What kinds of questions are you asking to which students? Who is sharing their work? Who is commenting on other students' ideas or representations?

Taking a Next Step

Identify three different students who come to your class with different strengths and experiences. Write a paragraph about each, including a goal for them in terms of their engagement with mathematics and with the mathematics classroom community.

Video 7.1 Transcript: **"She Wrote What It Meant"**

Mr. Fairley-Pittman [reveals Kedly's statement on the board]:	Can you read it for us, Jamillz, really loud?
Jamillz reads Kedly's statement:	When I add 1 to the first factor, the product increases by how much we're multiplying by.
Student:	What do you mean by "increases by how much we're multiplying by"?
Kedly:	I said that because in 7×3, you're multiplying 3 seven times. And since 7 increased by one group, since 7 increased by 1, that means the product increased by one group of 3.
Student(s):	Oh!
Esmerelda:	I get it. I get it
Mr. Fairley-Pittman:	What do you get?
Esmerelda:	I get it. I thought when he was talking about multiplying 7 three times, then when he explained it, I got it more because when he was saying you add 1 to 7 to get to 8, but . . . [she trails off; unintelligible]
Mr. Fairley-Pittman:	Hold on a second, I don't . . . What did Kedly say? Did you all understand what Kedly said?
Some students:	No.
Mr. Fairley-Pittman:	So maybe you could go up there and tell us what you meant one more time, Kedly? And let's listen really closely to what he's saying. You can use the markers, too, you can draw lines or anything . . .
Kedly [goes to the board]:	What I mean is that in the first problem, we're multiplying 3 and then in the second problem, 7 increased by one group to 8, so that means we added one extra group of 3 to get us to 24.
Alfred:	So the higher the factor goes, the higher the product will go.
Mr. Fairley-Pittman:	Are you saying, Alfred, that if we're increasing the factors, the product has to increase?
Alfred:	Yeah. [He then says something about another problem—indecipherable]

Mr. Fairley-Pittman:	Let's keep going. Let's keep looking at some of these. [Reveals Rollie's statement on the board.] Would someone care to read Rollie's?
Ariana reads Rollie's statement**:**	When I add 1 to the first factor, the product increases by the second.
Rollie calls on Kedly.	
Kedly:	I have a question. You said, "The product increases by the second." By the second what?
Rollie:	The second factor . . .
Mr. Fairley-Pittman:	You know that's a really interesting point because if we look at Hailey's [reveals Hailey's on the board] . . .
Students:	Say what? [and some other exclamations]
Mr. Fairley-Pittman:	Say what? Hailey, read yours really loud. Hailey wrote this.
Hailey:	When I add 1 to the first factor, the product increases by 1 of the second factor.
Mr. Fairley-Pittman:	Let's just think about Rollie's and Hailey's statements that they wrote right now. Why is everyone so surprised right now? I think I know, but I want to ask. Why is everyone so surprised?
Jaslanie:	They wrote the same thing.
Mr. Fairley-Pittman:	They wrote . . . did they write the same thing? [Students: no] No. It's similar.
Jaslanie:	Rollie wrote "second," but then Hailey, she wrote like what it meant.
Mr. Fairley-Pittman:	What do you mean "what it meant"?
Jaslanie:	Well, like, in Rollie's problem, when Kedly asked him what does it mean, he said the same thing, so that's what I think. So Hailey wrote like what "second" means, but Rollie didn't.
Kevin:	Well, I have a question for Hailey. What do you mean by "the product increases by 1 of the second factor?"
Hailey:	So, like, since we're increasing the first factor by 1, the product is going to increase by 1 of the second factor, because we're only adding one more group to the product.

Mr. Fairley-Pittman: You know what I think is making this a little bit challenging. Actually, there's two things I want to say. Number one: I think everyone was blown away between Rollie and Hailey, not only because they wrote very similar things, but also because the question that Kedly was asking was answered by Hailey's statement, right? You were asking what do you mean by "second," and what you meant, Rollie, was "second factor," right? So I think it's important to remember when we're writing these statements and conjectures that language and the words that we use, and being as precise as possible while also being kind of general, is really important. That one word right here, did this one word help your understanding, Kedly? [Yes.] Just one word changed your whole understanding of what's being written here. So we're going to try this again today, and I want you to think about that.

Student-Created Representations Offer Anchors, Openings, and Depth

Here are two student-created representations you saw in Chapters 1 and 3, respectively.

Figure P3.1 • A First Grader's Work Representing $3 + 6 = 9$, $9 - 3 = 6$, and $9 - 6 = 3$

1. Draw a representation for $3 + 6 = 9$.

2. Draw a representation for $9 - 3 = 6$

3. Draw a representation for $9 - 6 = 3$

Figure P3.2 • A Third Grader's Representations for Two Related Story Problems

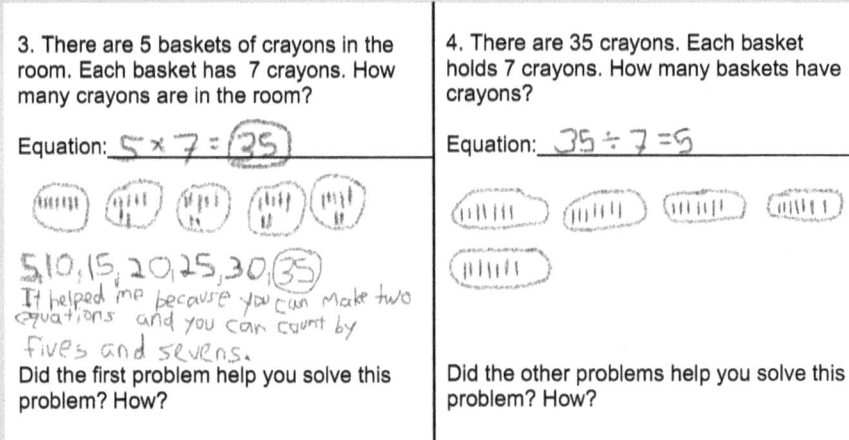

3. There are 5 baskets of crayons in the room. Each basket has 7 crayons. How many crayons are in the room?

Equation: 5 × 7 = (35)

510, 15, 20, 25, 30, (35)
It helped me because you can make two equations and you can count by fives and sevens.

Did the first problem help you solve this problem? How?

4. There are 35 crayons. Each basket holds 7 crayons. How many baskets have crayons?

Equation: 35 ÷ 7 = 5

Did the other problems help you solve this problem? How?

In Part One, where these representations first appeared, our focus was on how different modes of participation create openings for different students to participate in the class discussion—students speak or point, use many words or few, make observations about the equations or about different parts of drawings or models. Central to these openings is the use of student-created representations to focus the discussion. Such representations act as anchors for student dialogue, help everyone in the class express their ideas, enrich the discussion by avoiding an overreliance on only verbal communication, and communicate to all students how their own pictures and models can help everyone think about the mathematics.

In Part Three, we will take a closer look at how student-created representations play a central role in supporting both deep learning of mathematics and equitable participation. A key aspect of such representations is that they connect students to the meaning of the operations. With these images, students can see not just symbol patterns but can come to recognize *how* and *why* the operations work the way they do. Such understanding will be key as students make sense of calculation procedures and, in years to come, work with algebraic expressions and equations. Further, by making students' thinking visible, representations allow more students to follow and contribute to the discourse. Students can think through, clarify, explain, justify, and raise questions by referring to parts of the representations. Students have a better chance of following and understanding each other's thinking if it does not rely exclusively on words and equations but is also grounded in drawings and diagrams.

Representation is central to mathematical investigation at all levels as a tool both for exploration and for explanation of ideas. But representations such as pictures, diagrams, and manipulatives are sometimes used in elementary classrooms only briefly as a stage leading toward efficient use of symbols. For example, early in their study of addition and subtraction, students might be given physical objects to combine or separate until they can solve problems on paper, but then they are quickly weaned off manipulatives and drawings in favor of calculations with numerals. Similarly, when they are introduced to multiplication and division, students might again be shown drawings that represent equal-sized groups, but they are expected to quickly leave those representations behind and become fluent with facts and algorithms.

Many teachers have been taught that there's a strict progression from working with concrete materials to pictorial representations to working with symbols, which are referred to as *abstract*. We have found in the classrooms we've worked with over many years that thinking about a strict linear sequence from concrete to pictorial to abstract is problematic for two reasons: (1) It is used to sort students, preventing students labeled as "concrete thinkers" from engaging in higher-level thinking, and (2) it denies all students access to the powerful affordances of images and representations. Throughout students' study of mathematics, deepening understanding involves making connections among a variety of representations—physical models, drawings and diagrams, symbols, and words—as well as moving back and forth between images and abstract concepts. We have been struck, again and again, how students—even in the primary grades—articulate abstract ideas about the operations based on what they've noticed about symbol patterns, and then continue to construct and justify their ideas through representing these abstractions with manipulatives, pictures, diagrams, and story contexts.

Our focus in Part Three is on *student-created* representations, which include pictures and drawings, mathematical diagrams, physical models, and story contexts. Creating a representation is an *act of thinking through mathematical relationships and structures*: How is the 9 related to the 6 and the 3? How do I show which quantities are added and which are subtracted? Student-created representations are based on many elements that students incorporate into their mathematical repertoire as they develop their own mathematical voices. In these chapters, you'll see representations students created based on their own experiences with actions such as counting, joining, separating, and comparing quantities and on images of everyday or imaginary story contexts. You will also see students use forms of representation, like number lines or arrays, that they have learned about in mathematics class. The creation and analysis of representations both opens access to important mathematics and is an avenue for students' development of agency as mathematics learners.

The following are the major themes of Part Three:

- Representations in the form of diagrams, pictures, arrangements of physical objects, and story contexts support deep understanding of the mathematics. Creating, comparing, and critiquing representations such as these, as well as using equations and verbal expressions, help all students investigate the mathematical ideas, make meaning for symbols, and explain why a generalization is true.

- Student-created representations allow students who have different ways of thinking about the mathematics to express and explain their ideas.

- During class discussion, visual representations expand modes of participation and ground students' voices. Participation doesn't depend only on oral/aural facility.

- Focusing attention on student-created representations honors students' contributions to the collective development of ideas.

Chapter 8

Raising Student Voices Through Student-Created Representations

In Chapter 3, Isabel Schooler's first graders were considering the relationship between addition and subtraction, using representations of $7 + 5 = 12$ and $12 - 5 = 7$ (see Figure 8.1).

Figure 8.1 • Two Related Story Problems

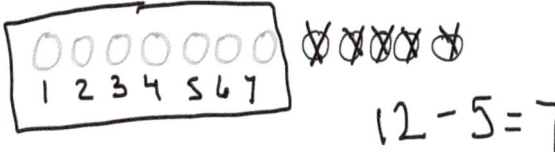

Jan is holding 7 red balloons, and Tim gives her 5 blue balloons. How many balloons is she holding now?

○ ○ ○ ○ ○ ○ ○ ○ ○ ○ ○ ○
1 2 3 4 5 6 7 8 9 10 11 12

$7 + 5 = 12$

Jan is holding 12 balloons. 5 balloons fly away. How many balloons is she holding now?

○ ○ ○ ○ ○ ○ ○ ⊘ ⊘ ⊘ ⊘ ⊘
1 2 3 4 5 6 7

$12 - 5 = 7$

Jeuri refers to the poster, which Ms. Schooler has re-created from a student's work, that shows 7 circles combined with 5 circles for the first problem and 12 circles with 5 crossed out for the second problem:

Jeuri:	It's minus 5 and then it was 7 left so it was . . . 12 minus 5 equals 7.
Ms. Schooler:	Now what's the same about that as what was in the first problem, Jeuri?
Jeuri:	You mean this? [points to the picture of 7 circles combined with 5 circles]
Ms. Schooler:	Yes!
Jeuri:	Oh. So it was 7 and then it makes it together [shows hands apart then claps hands together].
Ms. Schooler:	So the first problem is together? What do you mean?
Jeuri:	I mean adding.

Student-created representations such as these circles and Xs representing addition and subtraction support both rigorous mathematics and more equitable student participation. Using representations to investigate generalizations about the number system helps students to focus directly on each of the operations, highlighting the unique set of properties and characteristics of each. In order to dig into what is really going on—why does the operation work this way?—students reason with manipulatives, drawings, diagrams, and story contexts that embody the meaning of the operation. By considering and comparing a variety of representations, students identify connections among arithmetic symbols, diagrams, story contexts, manipulatives, and words, thereby strengthening their understanding of the underlying structure of the mathematics.

In this chapter, you will

- consider the characteristics of an effective student-created representation,
- create your own representation and story context for a conjecture about equivalent addition expressions,
- analyze three representations from a second-grade class,
- watch a video of the second graders considering two of the representations, and
- reflect on commentary about the importance of representation by several of our Collaborating Teachers.

Representations in the form of drawings, diagrams, manipulatives, and story contexts are central to mathematical investigation at all levels as a tool for exploration and explanation of ideas. Through describing and critiquing both their own representations and representations created by their peers, students build a repertoire of images for investigating mathematics and come to understand the mathematics they are depicting more deeply. Further, encouraging and centering representations as an expected part of mathematical activity considerably expands the ways students can express their identity as mathematicians. As they develop a repertoire of drawings, diagrams, story contexts, and models with which they are comfortable and experienced, they are building multiple ways to have voice in classroom mathematics and strengthening their agency in communicating their ideas.

As you consider the examples in this chapter, we'd like you to think about three aspects of the principle that *student-created representations offer anchors, openings, and depth*. These aspects of raising student voices through the centering of student-created representations lay a strong foundation for interweaving rigorous mathematics and equitable participation.

1. Creating drawings, diagrams, models, and story contexts for examples of the class conjecture requires students to investigate the underlying structure of the mathematical relationships.

2. Teachers choose representations to share that offer different ways into the mathematics so that every student can both find images that match their own ways of thinking and consider new ways of picturing the mathematical relationships.

3. Centering student-created representations honors students' thinking, provides a variety of openings into the mathematics, and communicates that students themselves have ways of illuminating the mathematics for everyone.

What Makes an Effective Representation?

Before analyzing some student-created representations and viewing a video in which such representations are discussed, we want to pause briefly to focus on the question, "What makes an effective mathematical representation?" How do you help students begin to understand what it means to represent a mathematical situation? How do you recognize student-created representations that will be effective tools for reasoning? Because this book focuses on how students learn about the arithmetic operations, this section addresses effective ways to represent the operations.

Effective representations embody both the quantities involved and the meaning of the operation. That is, they show both the relationships between the numbers in an equation and the action of the operation. Consider the difference between the two images in Figure 8.2 drawn by students to represent this story problem: Jan is holding 12 balloons and 5 balloons fly away. How many balloons is she holding now?

Figure 8.2 • Top example: A representation for 12 − 5 = 7 That Does Not Show the Relationship Between the Three Quantities or the Action of Subtraction. Bottom Example: An Effective Representation for 12 − 5 = 7.

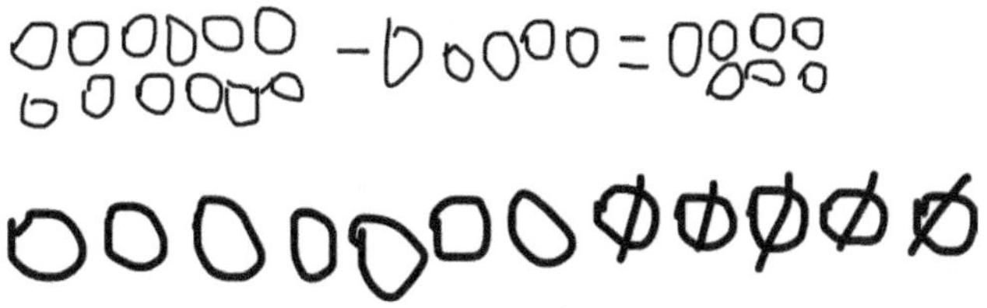

In the first representation, the numbers 12, 5, and 7 are shown as independent quantities. Subtraction is indicated by the minus sign, but the representation does not convey what subtraction means. It is as if the student has simply substituted a group of circles for each number in the equation 12 − 5 = 7. In the second representation, the meaning of subtraction is represented as removal. Twelve circles represent the 12 balloons initially held by Jan, and 5 of them are crossed out to show the 5 balloons that flew away. Seven balloons remain. The action of the operation is embodied in the second representation, and we can see how the 12 is composed of 7 and 5.

Young students don't automatically know what it means to "represent" a story problem or equation. Many teachers have told us that at the beginning of the year, when asked to represent an equation, some students will actually build the equation itself with blocks as in Figure 8.3.

Figure 8.3 • Building an Equation With Cubes

These students are not yet understanding that "to represent" the arithmetic in the equation is to show the action of subtraction and how the 12 can be seen as composed of 5 and 7.

Creating effective representations is a skill that is learned through multiple experiences. When first asked to represent a story problem, young students might simply draw a scene, maybe a picture of balloons flying away without attention to the quantities involved in the problem. But gradually, by sharing and discussing student work, students find their own ways to construct representations that show the quantities and the action of the operation (see Figure 8.4).

Figure 8.4 • A Representation of 5 Balloons Flying Away, Leaving 7

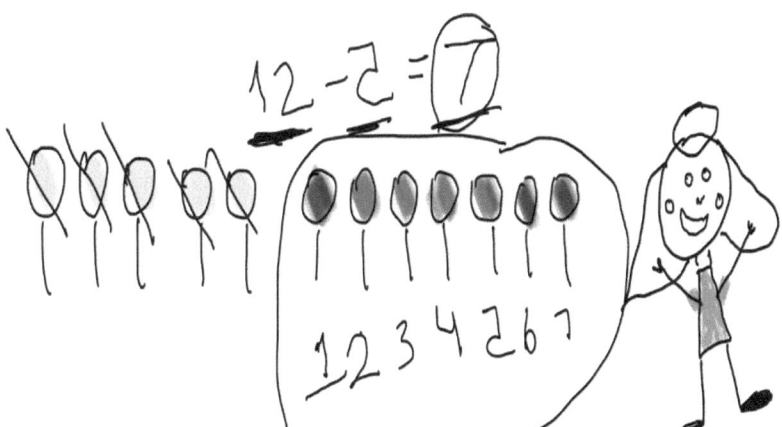

One way that teachers encourage students to develop representations that show the action of the operation and the quantities involved in relation to each other is by asking what we call "core questions." This series of questions asks students to identify how and where each quantity and each operation is shown in the representation. Ms. Gordon asks these questions in the Chapter 1 discussion of the representations of $3 + 6 = 9$, $9 - 6 = 3$, and $9 - 3 = 6$. She helps students connect the drawings with the equations by asking questions like, "Where do you see the 3 in the drawing? What does this 3 represent? How can you tell that this one is taking away 3?" For the story about Jan and her balloons, teachers would ask the following questions:

- Where do you see the 12 balloons Jan was holding at the beginning?
- Where do you see the balloons that flew away?
- Where do you see the balloons she has left?
- How does this picture show subtraction?

As you watch students create and describe representations in this and subsequent chapters, notice how teachers help them to connect their drawings or diagrams to story contexts and to equations. What kinds of questions do they ask in order to help students focus on the meaning and action of the operation?

Do the Math, Part 1: Investigate Representations for Equivalent Addition Expressions

In Chapter 4, you worked on noticing patterns and writing a conjecture about equivalent addition expressions like $10 + 5 = 9 + 6$ and $18 + 18 = 19 + 17$. In this chapter, we'll again visit Quayisha Clarke's second-grade class working on this idea (this is a different year with different students). In this class, students are working with the example $46 + 98 = 44 + 100$.

1. If you have your math work from Chapter 4, look back at the statement you wrote about equivalent addition expressions. How does it apply to $46 + 98 = 44 + 100$? Do you want to modify it in any way? If you don't have your Chapter 4 statement, write a new conjecture about equivalent addition expressions.

2. Create a drawing or diagram or a physical model (using cubes or base ten blocks, for example) to show $46 + 98$.

3. Show how your drawing or model can be changed just a little to show $44 + 100$. Use your drawing or model to explain why the sum doesn't change.

4. Write a story problem for $46 + 98$. Add another sentence or two to the story in order to illustrate $44 + 100$. Use your story to explain why the sum doesn't change.

Compare your drawing and your story. Share your drawing and your story with colleagues.

Do the Math, Part 2: Students' Representations for Equivalent Addition Expressions

Ms. Clarke's second graders have been working with equivalent addition expressions. They have a class conjecture, "If you increase an addend by any number and you decrease the other addend by the same number, then the sum will stay the same." They have also been thinking about how to use equivalent addition expressions to solve problems, for example, how can you create an equivalent problem for $46 + 98$ that is easier to solve? And how do you know the sum will be the same? Students have been creating their own representations to answer these questions. Many students have shown how $46 + 98 = 44 + 100$. Ms. Clarke has chosen three of these representations to share with the class.

Livia created a model with connecting cubes. She used one color for the 98 and another color for the 46. A re-creation of three stages of her model is shown in Figure 8.5. Albert made a drawing (see Figure 8.6). Uchenna created a story context (see Figure 8.7).

Figure 8.5 • Three Phases of Livia's Model Constructed With Connecting Cubes

a

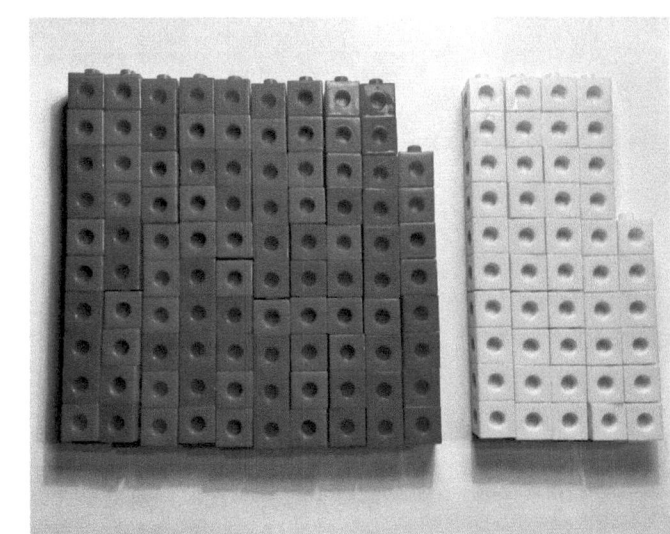

b

(Continued)

Figure 8.5 (Continued)

c

Figure 8.6 • A Re-creation of Albert's Drawing for
$46 + 98 = 44 + 100$

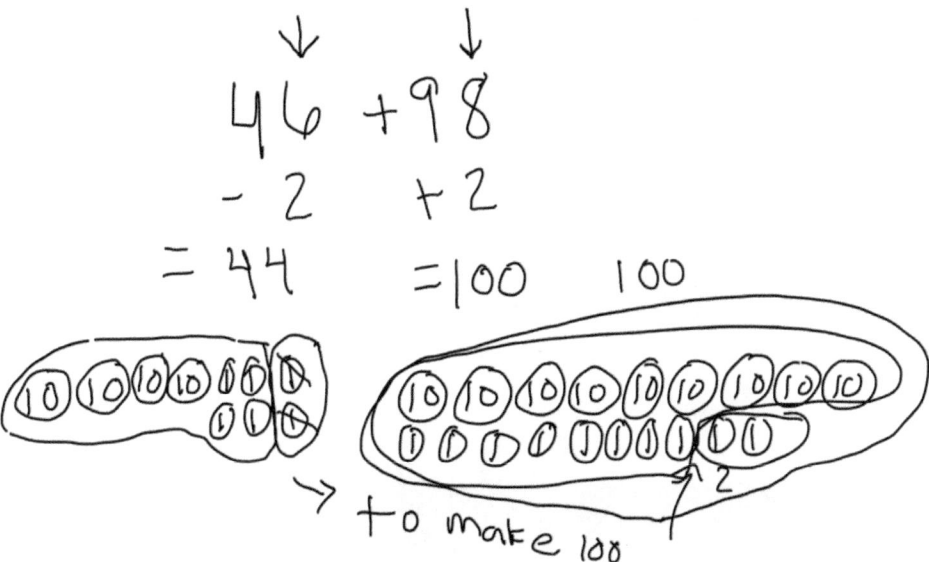

Figure 8.7 • Uchenna's Story for 46 + 98 = 44 + 100

46 + 98   Uchenna

                  100 + 44

Uchenna had 46 cubes Kylie had 8
cubes I dropped 1 cubes then Kylie picked
it up now the eoution is 45 + 99
we Started walking and Uchenna
dropped another one Kylie picked it up
now the eqoaution is 44 + 100. The
sum stayed the same because
Uchenna never lost any cubes Kylie
always picked them up.

46 − 1   45 − 1   44 + 100

98 + 1 = 99   99 + 1 = 100

> ## Reflection Questions: Characteristics of an Effective Representation in the Student Work
>
> 1. How does Livia's model show the different quantities 46, 98, 44, and 100? How does it show the change of 46 + 98 to 44 + 100? How does it show the sum of the two quantities?
>
> 2. How does Livia's model show an example of the class's conjecture, "If you increase an addend by any number and you decrease the other addend by the same number, then the sum will stay the same"? How can it be used to show that the sum is the same for the two expressions?
>
> 3. How does Albert's drawing show the different quantities 46, 98, 44, and 100? How does it show the change of 46 + 98 to 44 + 100? How does it show the sum of 46 + 98 and the sum of 44 + 100?
>
> 4. How does Albert's drawing show an example of the class's conjecture? How can it be used to show that the sum is the same for the two expressions?
>
> 5. How does Uchenna's story show the different quantities 46, 98, 44, and 100? How does it show the change of 46 + 98 to 44 + 100? How does it show the sum of 46 + 98 and the sum of 44 + 100?
>
> 6. How does Uchenna's story show an example of the class's conjecture? How can it be used to show that the sum is the same for the two expressions?

Selecting Student Representations for Discussion

Each of the second graders represented $46 + 98 = 44 + 100$ in their own way, and each representation highlights different mathematical elements. How does a teacher select student representations for a class discussion, taking into account the dual commitment to equitable participation and significant mathematics content? Here are three basic principles to keep in mind while making these choices.

1. Rotate whose work is presented. Perhaps there is a student who has been quiet lately and needs drawing out. Perhaps one student's work hasn't been featured for a few weeks.

2. Include at least one piece of work that is likely to be accessible to all or most students.

3. Make sure the collection illustrates different mathematical features.

Think about the pieces selected by Ms. Clarke. Livia's cube model shows the base ten structure of the numbers and shows the subtraction and addition of 2 dynamically as the movement of two cubes from the 46 to the 98. Her use of different colors for the original addends of 46 and 98 enables viewers of her representation to track the movement of the two cubes. The sums of $46 + 98$ and of $44 + 100$ are seen as the joining of the two sets of cubes.

Albert's diagram also draws on a base ten model of the numbers. Albert's drawing assigns relative values by labeling each circle as either a 10 or a 1, similar to a model using dimes and pennies. He crosses out two of the circles and, with an arrow, shows that they are taken away from the 46 and joined to the 98. The total of 144, seen as the two sets of 10s and 1s together, does not change as the two 1s are shifted.

Uchenna's story embodies the addition and subtraction as picking up and dropping cubes. The sum of the two addends in the conjecture is Uchenna's and Kylie's cubes together. You might not think of a story as a "representation," but we have found that, in the elementary grades, stories are a particularly effective way for students to show and explain the mathematics as well as to illuminate other representations such as drawings or diagrams. Simple stories like Uchenna's, based on familiar experiences and actions, create strong and lasting images that students are able to call on. Story contexts don't need to be elaborate. In fact, they are often most effective when they convey a straightforward situation that embodies the action of an operation—dropping and picking up, comparing, accumulating equal groups, dividing some quantity into equal groups, and so forth. Student-created story contexts are based on students' real experiences and interests—they might be realistic, but they often have an element of humor or fantasy. In one third-grade class, students were studying equivalent subtraction expressions (which you encountered in Ms. Clarke's class in Chapter 5). One student came up with a story context about a boat that could only hold 12 people without sinking; if 13 people were on the boat, 1 had to jump off (13 − 1); if 14 people were on the boat, 2 had to jump off (14 − 2), and so forth. In this way, they generated a sequence of equivalent subtraction expressions. As one student said, "Every time another extra person gets on, you have to increase the number of people that jump off." This image stuck with other students in the class, and they referred back to it as they developed their conjectures and representations about equivalent subtraction expressions.

Livia, Albert, and Uchenna each created an effective representation that includes the quantities and operations needed to illustrate how 46 + 98 is equivalent to 44 + 100. Each provides different ways in for different students. As we continue to focus on representations in Part Three, keep in mind that a student-created representation is not only the static drawing or model or story the student has made. A complete representation may also include students' words, gestures, and actions used to demonstrate and explain. If we were watching Livia talk about her model, with the actual cubes in front of her, she could show us how she moves the two cubes from one set to the other. And Uchenna's story could be acted out; in fact, after Uchenna read his story to the class, Ms. Clarke showed with her hands the actions of dropping and picking up the cubes.

Watch the Video: "How Does This Show Addition?"

In this video clip, Ms. Clarke first projects photographs of different stages of Livia's model, then projects Albert's drawing for the class to discuss. As you watch Video 8.1, think about how centering student-created representations enhances students' learning and creates openings for students' voices. This clip shows only a brief example of student interaction, but the following questions are important to keep in mind as we continue the focus on representation in the next two chapters.

Video 8.1

"How Does This Show Addition?"

qrs.ly/6gfs4ye

Reflecting on the Video:
The Role of Representations in Supporting
Students' Identity and Agency

[You may want to use the transcript at the end of this chapter as you consider these questions.]

1. Look at each of the questions Ms. Clarke asks during the discussion. What do you think the purpose of each question is? What might be its effect?

2. Why might the sharing of their representations be important for the students who created them, in this case, Livia and Albert?

3. Imagine if Livia and Albert were explaining their ideas in words, without their representations. What might be different about the discussion in terms of student access and participation?

4. How does having students *create* their own representations support their identity and agency as mathematicians?

5. What is the potential of *sharing* student-created representations to support students' identity and agency as mathematicians—both for the students whose work is shared and for the rest of the class?

Read and Reflect on Collaborating Teachers' Comments about Student-Created Representations

Let's return to the three aspects of the principle that *student-created representations offer anchors, openings, and depth* listed at the beginning of this chapter.

---

1. Creating drawings, diagrams, models, and story contexts for examples of the class conjecture requires students to investigate the underlying structure of the mathematical relationships.

2. Teachers choose representations to share that offer different ways into the mathematics so that every student can both find images that match their own ways of thinking and consider new ways of picturing the mathematical relationships.

3. Centering student-created representations honors students' thinking, provides a variety of openings into the mathematics, and communicates that students themselves have ways of illuminating the mathematics for everyone.

Our Collaborating Teachers commented on two major aspects of using student-created representations in their classes: (1) how the representations drew students' attention to mathematical structure, and (2) how creating and sharing representations contributes to equitable participation.

1. Supporting Mathematical Thinking With Student-Created Representations

Jeff Parks: When I first started working with representations, I hadn't thought of them as a problem-solving tool. I eventually came to see that just as you can revise your language, you can also do that with representation—it's like a flow, it's an evolving thing.

So many students think of math as the equation. Every year, when I first ask students to represent an addition equation, someone builds the equation with blocks [see Figure 8.3]. Even some of my most typically successful students who know the answers will do this. It makes me realize that starting with the equation, with the abstract, too soon disadvantages the students in understanding why these symbols were developed. They don't realize the equations represent something in the world. I want to start with, what is the problem you have to solve, what are you looking for, what are the values, what is the action of the operation? Using a representation highlights that. When you move to the abstract too soon, you lose the thinking and the process. The connection between equations and context has deepened for my students, and they are beginning to understand that mathematical notation is just one way to represent the world.

Michelle Sirois: A few years ago, when I taught the strategy of creating an equivalent expression to solve an addition problem, like making $198 + 65$ into $200 + 63$, I moved through the idea quickly. I was thinking that's at a higher level—some kids are going to get it but not everyone. Once I started using representations, I taught that lesson differently. I taught it around the question, What story context can we think of right now to make sense of this idea? Many students use a story context to make sense of what's happening in other representations. The stories really anchor them and often, when they are trying to make sense of other visuals, they refer back to a story. Doing so allows them to make connections independently. One girl was into hair bows, and she read a story about having so many hair bows in one bucket and so many in another. And another student wrote about how he had so many chicken nuggets and his brother had so many chicken nuggets. They also came up with their own language—they said, "You're just *shifting* from one group to the other." So then when we got to solving problems for which you could make an equivalent expression, I could just say "It's like your chicken nugget problem," or "It's like your hair bows," and then they could apply the strategy. And then when we were working with mixed numbers and fractions, we were doing a problem, $1\frac{7}{8} + \frac{4}{8}$; I looked at a kid's work and he just had the answer $2\frac{3}{8}$, but he had no work. I made the assumption that he had just looked at his neighbor's work to get the answer. But I decided to ask him how he solved it. When I did, he said, "Oh, it's just like a shifting problem. I just took $\frac{1}{8}$ from the $\frac{4}{8}$, and made it two wholes, and then I had $2\frac{3}{8}$." That moment really taught me to check my assumptions, and it was also an example of how you transfer stories and representations to other work. It anchors them. I used to think, in fourth grade, this was a strategy only some kids could get, but, no, everyone can get this.

Quayisha Clarke: My class has discussed student-created representations before but only in reference to individual problems. Now, the purpose for representation is moving away from showing how to arrive at the answer to one particular calculation to showing how two problems are related. This shift requires students to see the two problems within the representation and to make sense of how the first problem helps us to quickly know the answer to the second problem.

2. Supporting the Mathematics Community Through Student-Created Representations

Isabel Schooler: Representations are where math lives. It's not just showing what *you* know and what *you* are thinking—your representation helps other students learn, too. Highlighting certain representations brings students' thinking forward. It brings the work to life, especially for learners who are more visual. A lot of school is conversation, and if we can have a conversation around something visual, that can help more students understand.

Sometimes I will highlight a representation that shows a more advanced strategy that we are working toward, and you'll see little "ah ha" moments, "Oh, I get it now."

I think also, with equity of voice, you may choose a representation from a student who doesn't speak up as much. That's on the teacher to keep track of. Whose work has been shown? Whose voices have we heard?

Emmanuel Fairley-Pittman: The representations are a tool, not only to get to the right answer but also to get someone to understand what you're thinking. Without representations, I don't think you can meaningfully engage all students in the conversation. You can have conversations, but there are students who don't know what's going on. Representations allow students to talk about the math, think about it more meaningfully, and develop a deeper understanding. We have to have representations, or we can't have these conversations.

The hard part is making time for analyzing someone's representation. It's not always going to be pretty, it's not always going to be beautifully drawn. You do have to wrestle with it, and that's okay. That's what it means to be a learner in a shared space, making time and space to wrestle with someone else's ideas. Sharing your representation positions you to be the teacher, it pushes you to engage more collaboratively with the group. You're not creating a representation for yourself, you're creating it so someone else understands what you're thinking. If something is unclear about what you've written or drawn, you need to revise so it's more clear. The representation is about making things clear for other people.

Jeff Parks: My focus on student-created representations is also about equity. Every year in third grade, you have kids who come in and say, "This is the algorithm, this is how you do math," but they have no idea what they're doing. They just follow steps and run into lots of difficulties. By basing it in representation, students are making sense of situations that come up in their world. Also, for someone who may not be viewed as successful or who doesn't feel confident in their work, choosing their representations to share makes them feel connected. I have a student who has a lot of ideas, but she doesn't talk in class. I wanted to highlight her work and highlight her thinking and help her feel engaged with the lesson. Representation supports student confidence and agency. Students become more flexible. They know, "I can figure this out, I can use a representation with something I don't understand to problem-solve." They become OK with uncertainty. They are willing to take a step back and say I don't know YET, but I can figure this out.

Reflecting on Collaborating Teachers' Comments

1. Mr. Parks notes that "so many students think of math as the equation." What do you think he means, and do you notice that his observation is true in your own context? Why might "seeing math as the equation" be problematic?

2. Much of school relies on words—speaking them, reading them, and hearing them. How do you see this play out for different students in mathematics?

3. Ms. Schooler and Mr. Fairley-Pittman make strong statements about the necessity of using representations in order to support equity. What is your reaction to their statements? What thoughts do you have about the use of representations and equity in mathematics in your own context?

What Do You Want to Remember From This Chapter?

Take a few minutes to note for yourself ideas you want to hold onto as you continue to investigate how the use of student-created representations supports the building of a mathematics community. What teacher moves have you noticed in this chapter that you want to bring into your own practice? Here are some of the thoughts we have had as we listened to our Collaborating Teachers and considered the work of their students:

- **Encourage and support *a variety* of student-generated representations:** Diagrams, pictures, and story contexts help students develop a strong sense of what an operation means. Know that representations don't have to be the same, and they don't have to look like the ones in your instructional materials or look the same as you would create. What matters is that students can use their own and classmates' representations to deepen their understanding of the properties and characteristics of different operations and make sense of why each operation requires different steps in calculations.

- **Encourage story contexts.** Create your own story contexts as examples, and highlight story contexts created by students that show the actions of the operations. Encourage use of story contexts to explain drawings, models, and equations. Pay attention to story contexts that seem to "stick" with students; remind students to use these images.

- **Rotate whose work is presented.** Over time, make sure all students have opportunities to share their representations. Think about Ms. Schooler's questions, "Whose work has been shown? Whose voices have we heard?"
- **Ask "core questions" about each representation.** Ask students where they see each quantity in the representation and how they see the operations. Asking the same questions about different representations allows students to see correspondences across representations and make deeper connections.

Taking a Next Step

Consider your own classroom or talk with a teacher about their class: List all the ways representations have been used in mathematics class over the past two weeks. When did you use representations to demonstrate a concept? When did students encounter representations in their math textbook or worksheets? When did students create their own representations? In any of the instances you list, were students considering a single representation or multiple representations of the same idea? What evidence do you have that students used the representations to think about the mathematics more deeply? What evidence do you have that the use of representations gave voice to students, and to which students?

Look over your responses. What do you notice? Are there implications for your practice?

Video 8.1 Transcript: "How Does This Show Addition?"

Ms. Clarke:	Now we're going to go over the same questions for each representation. And I want you to be critical, OK? Mathematicians, even though we're showing someone's work, it's really important that you're honest with what you see and what you think because that's going to help the mathematicians in the room and it's going to help me. OK? So let's take a look. Our first representation might look similar to many of your representations. We had a couple of people working with cubes over the last two days. Now, Livia said that she would want to share her representation. Our first expression was 46 + 98 [records this on the whiteboard], and we wanted to see if we could make an equivalent expression that's easier to solve. So I want you to take a look, and, Livia, can you just explain to us what you did first and what you did second?
Livia:	First I put 98 and [then] 46, and then I added 2 to the 98, and then, from the 46, from the 46 it has, now it has 44.
Ms. Clarke:	So, with her representation, these were her addends, and then what are you showing here?
Livia:	How many it had altogether.
Ms. Clarke:	And then we're taking a look at what she did. And what does this show, Livia?
Livia:	What I got.
Ms. Clarke:	OK, so I want you to think about, how do you see addition in her images and in what she said? How does this show addition?
	Ms. Clarke now has students turn and talk about these questions for about 45 seconds. We rejoin the class at the end of the turn-and-talk.
Ms. Clarke:	How does this representation show addition? Anu, what do you think?
Anu:	Because she added the black ones to the yellow ones.
Ms. Clarke:	And how do you see that in the representation? Josiah?
Josiah:	We see that because, we see the two blacks that were on, with the other ones, are on the yellow.
Ms. Clarke:	And is that what you were saying, Anu? You just said we see them showing addition how?
Anu:	Because she added the black one to the yellow one.
Ms. Clarke:	Which cubes were you talking about? Which black cubes?
Anu:	Those black cubes.

Ms. Clarke:	Can you come up and point cause I want to make sure we're all talking about the same thing. [Anu comes up and points.] OK, so, Josiah, that's what you were talking about as well? Here we can see black cubes [points to the group of 98 black cubes in Figure 8.5a], no black cubes [points to the group of 46 yellow cubes], and then she took 2 and put it over here [pointing to the 2 moved black cubes in Figure 8.5b]. So you think that shows addition? Is there any other way that you see addition shown in this representation? Joniah?
Joniah:	Right there.
Ms. Clarke:	Can you show us and explain your thinking?
Joniah:	Because there's a black cube, and she put it together.
Ms. Clarke:	Can you show us where she put it together? [Joniah points to Figure 8.5c.] Okay. So what do you guys think about that? What do you think about that?
Angelina:	I agree with Joniah.
Ms. Clarke:	OK, can someone add on?
Josiah:	I agree with Joniah because she's adding; she added the 2, but then she added the rest of them to the 100. When she added the 2 more, it equaled 100.
Ms. Clarke:	So she's putting her addends together. So what does that give us, when we put our addends together?
Tunmishe:	It tells us the total.
Ms. Clarke:	It tells us the total. And another word for the total? Uchenna?
Uchenna:	The sum.
Ms. Clarke:	The sum. OK. I just want to make sure we're really looking at all of the work here.
Ms. Clarke:	This is Albert's representation [Figure 8.6]. Albert, can you explain what you did here?
Albert:	So I knew that if I took away 2 from the 46, I would get 100. So I made the 10 and then the 1s to show that there was only 44 left, and I added the 2 to 98 [unintelligible] 100 and when I crossed out those 2 from the 46, I also [unintelligible] the 2 in the 100 to show that those 2 were the ones that were added to the 98.
Ms. Clarke:	So looking at Albert's representation and thinking about what he said, how does this representation show addition? How does this representation show addition? Uchenna?
Uchenna:	It shows addition over here because when he took away that, he added it to the 98 and that equaled the 100, and because he added right here, too, he added that 100 to those ones, and I can see that because there's a plus sign.

Chapter 9

Looking Across Representations

As students become more practiced in creating and sharing representations about the patterns they've noticed, they can dig more deeply into what those representations offer them. They analyze, critique, and compare representations and think about how those representations embody their conjectures. In this chapter, we examine what students gain mathematically as they analyze two student representations of the same problem, think about how the representations connect to their conjecture about adding 1 to an addend, and notice how different parts of their conjecture are represented differently. At the same time, we'll look at how sharing and comparing representations is a catalyst for student participation. First, when a student's work is presented for deep analysis, that student is validated as a mathematical thinker with ideas that are worthy of the time and attention of their classmates. Second, the variety of representations provides multiple entry points, giving greater access to engagement in the discussion.

> **In this chapter, you will**
>
> - analyze two first graders' representations of a pair of related story problems,
>
> - analyze a video of a first-grade lesson in which the class discusses the same two student-created representations, and
>
> - consider the reflections of our Critical Friends on how students look across representations to gain deeper understanding of the mathematics.

As you do these activities, we'd like you to think about these three aspects of the principle that *student-created representations offer anchors, openings, and depth.*

1. Looking at and comparing several students' representations helps all students deepen their understanding and make meaning for symbols.

2. Each student is expected to make sense of other students' mathematical ideas as embodied in their representations.

3. Critical attention to a student's representation recognizes that student as having important ideas to communicate.

Do the Math

You are about to view clips from Natasha Gordon's first-grade class, whom you first met in Chapter 1. The lesson takes place in the middle of a sequence of lessons in which students are investigating what happens to the sum when one addend increases by 1. The sequence began by asking students what they notice when given pairs of equations like these:

$6 + 10 = 16$ $6 + 11 = 17$	$6 + 10 = 16$ $7 + 10 = 17$
$4 + 7 = 11$ $4 + 8 = 12$	$4 + 7 = 11$ $5 + 7 = 12$

Based on their observations, the class composed a conjecture:

If you add 1 to an addend, your total is going to increase by that 1 you added to the addend.

As they continued to investigate the conjecture, students worked individually to create representations for the following story presented by their teacher:

Ms. Gordon has 8 bottle caps. Ms. Schooler has 12 bottle cops. When they put their collections together, they have 20 bottle caps.

Ms. Schooler found another bottle cap. Ms. Gordon still has 8 bottle caps, but Ms. Schooler now has 13 bottle caps.

In preparation for the class discussion, Ms. Gordon selected two samples of student work to share, Livia's and Kaitlyn's, shown in Figures 9.1 and 9.2, each of which had different features for the class to interpret.

Figure 9.1 • Livia's Work

1. How many do they have together?

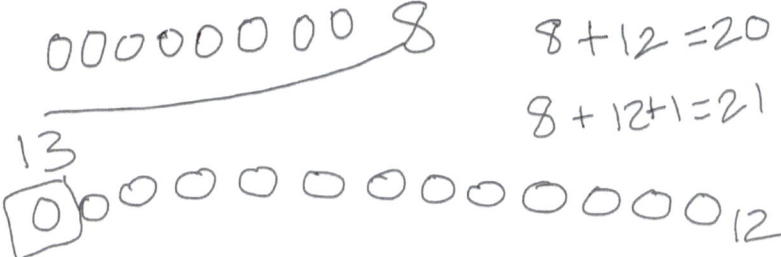

Figure 9.2 • Kaitlyn's Work

1. How many do they have together?

1. Consider the core questions for Livia's and Kaitlyn's representations:
   - Where do you see 8 and 12?
   - How does the representation show 8 and 12 are added?
   - Where do you see the sum?
   - Where do you see the 1 that was added to 12?
   - How does the representation show the sum increased by 1?

2. In what ways are Livia's and Kaitlyn's representations the same, and in what ways are they different?

3. Why is it important that students' analysis and discussion of these representations shift from an exclusive focus on quantities to include the action of the operation?

Watch the Video: "Where Is Ms. Schooler's Additional Bottle Cap?"

There are three video clips from this lesson: class discussion of Livia's work, class discussion of Kaitlyn's work, and class discussion of aspects of Livia's and Kaitlyn's work combined. We recommend that you watch each clip separately, considering the reflection questions for that clip before moving on to the next.

Viewing of Video 9.1: "Where Is Ms. Schooler's Additional Bottle Cap?" Part 1

Ms. Gordon has displayed on the whiteboard Livia's work (Figure 9.1). At first, Ms. Gordon covered up the equations, $8 + 12 = 20$ and $8 + 12 + 1 = 21$, and asked students to talk in pairs about what they notice about the student's drawing. The video begins with the final 15 seconds of their turn-and-talk.

Video 9.1

"Where Is Ms. Schooler's Additional Bottle Cap?" Part 1

qrs.ly/gufs4yf

Reflecting on the Video, Part 1

[You may want to use the transcript at the end of the chapter as you consider these questions.]

1. What connections are students making across different forms of representation—Livia's drawing, the equations she wrote, and the verbal statement of the conjecture?

2. How does a focus on Livia's drawing bring students into the discussion?

3. What purpose does it serve to first hide the equations and then, several minutes into the discussion, reveal the equations?

Viewing of Video 9.2: "Where Is Ms. Schooler's Additional Bottle Cap?" Part 2

Video 9.2

"Where Is Ms. Schooler's Additional Bottle Cap?" Part 2

qrs.ly/jhfs4yi

Reflecting on the Video, Part 2

[You may want to use the transcript at the end of the chapter as you consider these questions.]

1. What makes Ryan think Kaitlyn's drawing does not match the story?

2. Why might Ms. Gordon have invited Ryan to say what he found confusing in the representation?

Viewing of Video 9.3: "Where Is Ms. Schooler's Additional Bottle Cap?" Part 3

Video 9.3

"Where Is Ms. Schooler's Additional Bottle Cap?" Part 3

qrs.ly/xtfs4yq

Reflecting on the Video, Part 3

[You may want to use the transcript at the end of the chapter as you consider these questions.]

1. How do students make sense of the equation $8 + 12 + 1 = 20 + 1$?

2. What might be the purpose of Ms. Gordon's decision to write an equation that includes one expression from each student's work?

Read and Reflect on What Others See in the Video

Let's return to the three aspects of the principle that *student-created representations offer anchors, openings, and depth* listed at the beginning of this chapter:

1. Looking at and comparing several students' representations helps all students deepen their understanding and make meaning for symbols.

2. Each student is expected to make sense of other students' mathematical ideas as embodied in their representations.

3. Critical attention to a student's representation recognizes that student as having important ideas to communicate.

When our Critical Friends viewed the video clips "Where Is Ms. Schooler's Additional Bottle Cap?," their comments addressed these three aspects. They began the discussion with what Livia's representation offered the class.

1. Interpreting Livia's Representations

Imagine trying to lead a discussion about this conjecture without using a drawing, diagram, or manipulatives. If the discussion were just words, it would be hard to follow what each person is referring to. Which number are you talking about? Which addend? As students are working to find the words to convey what they see, Livia's drawing provides a reference, grounding the mathematical ideas in an image.

Lynne Godfrey: The drawing was really powerful for the class to be able to talk about the quantities and the action and what was happening in the problem. They used the student work to make sense of what that extra 1 was.

Darlene Ratliff: Toward the beginning of the discussion, the teacher asked, "Can you show us the quantities you're talking about?" So those sitting on the rug who might not have known how to interpret the drawing could see, this is the 12, this is the 1. And then after several students explained or pointed to the components of the problem, Ms. Gordon repeated it as well. If there were kids in the class who didn't initially understand what was happening, that was clearly presented so they could tap into the discussion.

Yi Law Chan: I've been thinking a lot about the importance of having students see connections between different modes of representations. I was noticing how this teacher kept asking the class to connect back to the story. They didn't just stay with the drawing. After the student swept his hand to show where Ms. Schooler's initial 12 bottle caps appeared in the diagram, the teacher said, "Right, because the problem said that she found another bottle cap, and now this is the extra 1." She didn't just say that 1 was added; she emphasized that it was added *because* Ms. Schooler found another bottle cap. The class kept moving between the visual representation and the story and eventually to the symbolic equations and their conjecture.

2. Different Ways of Looking at Kaitlyn's Representation

We each bring to a representation our own ways of seeing and making connections, and it may initially be hard to view it from another's lens until we listen hard to understand how they see it. Our Critical Friends initially had that experience listening to Ryan and appreciated the time Ms. Gordon spent drawing out Ryan's thinking.

Marta Garcia: Kaitlyn's representation emphasizes the change from 20 to 21. In the discussion of Livia's work, the students were more focused on the change from 12 to 13. Looking at both in a single lesson could help students hold onto the different parts of their conjecture.

Virginia Bastable: I looked at Kaitlyn's representation and said, well, she drew the first line of circles for the first problem, and then for Ms. Schooler's extra cap, she drew that on the next line. I had a perfectly good explanation for what Kaitlyn did. I wouldn't have guessed that it didn't make sense to some of the kids until I started listening to them. Livia had drawn a square to show the extra 1, and Kaitlyn showed the extra 1 by its position on the paper. But some students were confused by where on the paper Kaitlyn drew that 1.

Marta Garcia: I really loved the part of the lesson when Ryan says something is confusing. It is so important to be able to ask questions and express your own confusion. And he's given the time. Even though some kids' hands are waving as he's trying to talk, he keeps at it. He's trying to use the representation to bring clarity to the group about why he's confused.

The representation can help support students with the mathematics, but it can also support students when they're confused—they can use the representation to help unpack where their confusion is. The connections made across representations and contexts can also support multilingual students in making use of their linguistic resources. Moving back and forth among different types of representations (verbal, symbolic, physical, visual) supports multilingual students in both building their mathematical ideas and in participating in discourse practice.

Hetal Patel: After Ryan stated what confused him, Ms. Gordon asked other students to add on to what he said. It wasn't clear if they found Kaitlyn's representation confusing as well, but they had suggestions about how to revise her drawing to clarify that the 1 was added to Ms. Schooler's bottle caps.

3. Using Parts of Each Student's Work to Highlight a New Idea

The fact that Livia and Kaitlyn had written two different equations provided Ms. Gordon with the opportunity to take an expression from each in order to introduce a new equation: $8 + 12 + 1 = 20 + 1$. In many classrooms, when the teacher asks if such an equation makes sense, the answer would be a resounding No! That is, many students reject the possibility of having anything other than a single number to the right of the equal sign. Students often interpret "=" as an instruction to perform a calculation. For them, the equal sign means "The answer's coming up," so they would say that an equation like $3 + 4 = 4 + 3$ is simply wrong. If they have previously only seen equations in the form "$3 + 4 =$ ___" where they are expected to fill in "7," we can see where such an interpretation comes from. However, in Ms. Gordon's class, students had been learning the meaning of the equal sign as indicating equality and at least some of the students agreed that the equation $8 + 12 + 1 = 20 + 1$ did make sense.

Virginia Bastable: We know lots of kids have trouble with equations that have plus signs on both sides of the equal sign. Perhaps Ms. Gordon's class has had previous discussions about the meaning of the equal sign. It's still a question in my mind, why do they believe this? One student said, "Well, they don't look the same, but it has the same value." That was one student's idea of what equality is. But I don't know how common that belief was. Of the other students who spoke, it was hard to interpret what they meant.

Hetal Patel: I was also wondering if the discussion could have gone longer to go back to the story and the drawings to show why these two expressions are the same as one another.

Lynne Godfrey: It's the end of that particular math class, but it's December, and they still have many months left of school. So that didn't worry me. Maybe Ms. Gordon put the idea out there before everybody was ready to think about it. They will return to it again and again.

4. Each Student Is Expected to Make Sense

Making sense of mathematical ideas is not the same as listening carefully and repeating what was said. It's an internal process of making connections, often associated with a feeling of things clicking into place. Even though the students in a class are engaged in community, making sense is an individual process that doesn't happen for everyone at the same time or in the same way.

A part of making sense is to be able to know when something is *not* making sense *to you*, even if others are saying that it does. A step toward making sense is to try to say *what* doesn't make sense, as Ryan did in this clip.

Virginia Bastable: Throughout the lesson, often when students offered an idea, Ms. Gordon asked them to elaborate. "How so?" "What do you mean by that?" She holds the space open for them to continue thinking and presses for more.

Hetal Patel: I keep thinking in this discussion there's a value in making sense. The point is not just to listen; it's to understand. The students in this class are often explaining why. The teacher's questions have instilled in them that value.

Reflection Questions

1. The Critical Friends talk about how students are using their representations to better understand their conjecture. How do the representations make visible why the sum increases by 1 when an addend increases by 1? What is different about creating a representation that is an example of the students' conjecture rather than creating a representation to solve a calculation problem?

2. Virginia Bastable said, "I wouldn't have guessed that [Kaitlyn's drawing] didn't make sense to some of the kids until I started listening to them." What lesson might you take for your own practice from her comment?

3. Marta Garcia coins a phrase that we think is worth thinking about further. She says that Ryan is using the representation "to bring clarity to the group about why he's confused." How does it make sense that "clarity" and "confusion" are both at play in this lesson? In your own context, what do you think students and teachers believe about the role of confusion in mathematics learning?

What Do You Want to Remember From This Chapter?

Take a few minutes to note for yourself ideas you want to hold onto as you continue to investigate how the use of student-created representations supports individual and collective mathematical agency. Here are some of the thoughts we have as we consider the discussion among Ms. Gordon's students and the comments of our Critical Friends:

- **Present the work of more than one student.** Analyzing a variety of representations provides multiple entry points into the mathematics content. Make sure the collection illustrates different mathematical features. Making connections across different representations deepens students' understanding.

- **Consider carefully whose work to share.** When selecting student work to share with the class, consider both issues of equitable participation and mathematics content. Ask yourself such questions as, Who needs to be brought into the center of the discussion? Whose work hasn't been shared recently? What does each representation offer for student learning? Include at least one piece of work that is likely to be accessible to all or most students.

- **Ask students to interpret how each representation connects to their conjecture.** Student-created representations offer insights into the mathematical structure of the operations—how they behave and how they are different from each other. Once students are formulating conjectures that they think are true, representations provide access to understanding *why* they are true.

- **Encourage students to share their confusion.** Examining confusion is a step toward clarity. Part of making sense of mathematics is knowing when you're confused, figuring out what confuses you, and working collaboratively toward clarity.

Taking a Next Step

In Chapters 5 and 6, the Next Steps were to engage students in noticing patterns and writing conjectures about equivalent addition expressions. Post some examples of equivalent addition expressions they previously looked at as well as the students' conjectures, and revisit this work with the students. Ask students to come up with their own representations—drawings, arrangements of cubes, or story contexts—for a pair of equivalent addition expressions. As students are working on their representations, ask core questions about how their drawing or model shows each quantity and operation.

Video 9.1, 9.2, and 9.3 Transcripts: "Where Is Ms. Schooler's Additional Bottle Cap?" Parts 1, 2, and 3

9.1. "Where Is Ms. Schooler's Additional Bottle Cap?" Part 1: Discussion of Livia's Work

[15 seconds of students finishing turn-and-talk]

Ms. Gordon:	What did you discuss with your partner? What do you notice from the scholar's representation? And what does it tell you? Joniah?
Joniah:	I see 12 circles, but I see that 13 has a square so that means he added 1 more.
Ms. Gordon:	Oh, interesting. Josiah, you're saying you agree. Can you say more?
Josiah:	Like she has your 8 circles on the top, and then she has 12 circles, but the one that's the thirteenth one has a square.
Ms. Gordon:	Anu, you want to say more?
Anu:	It has a square because of so we know that that's what she added.
Ms. Gordon:	Huh, and you know I was going to ask you where do you see the different parts of the problem in the representation, but it seems like you already went right there. So, in the representation you can see Ms. Gordon's, you can see my 8 bottle caps, and from the original problem you can see the 12 bottle caps that Ms. Schooler had. Can someone come show everyone the 12 bottle caps just to make sure we're clear? Josiah said that you could see my 8 bottle caps on the top. Where do you see Ms. Schooler's initial 12 bottle caps? Alec?
Alec:	I think right there. The first ones right here.
Ms. Gordon:	Oooh okay, so these are her 12 initial bottle caps. Then what happened here, Anu? Why is this here?
Anu:	Because . . . [inaudible] she added it.
Ms. Gordon:	Then the problem tells us that Ms. Schooler found another bottle cap and now she has 13. So instead of showing a whole 'nother representation, drawing something completely different, what did you do, Livia?
Livia:	I drew a box.
Ms. Gordon:	Why?
Livia:	So people can know that one's the 1 that Ms. Schooler found.
Ms. Gordon:	Ah, to differentiate, to show you that this is the additional bottle

cap that she found, and this will now be her 13th bottle cap. Now we looked at the work that this scholar did to represent this problem. Now I would like us to look at . . . what do these equations tell us?

Josiah: It's like our conjecture because first it's 8 plus 12 equals 20. Now it's 8 plus 12 plus 1 equals 21. And we're adding 1 more, and our total is getting bigger. And our total is getting bigger.

Ms. Gordon: Can someone say more? It's like our conjecture, we're adding 1 more, our total is getting bigger. Can someone say more? How is it like our conjecture? Anu?

Anu: Because our conjecture says that if you add one number to an addend, the number's gonna increase by . . . , the addend's gonna increase by the number you add.

9.2. "Where Is Ms. Schooler's Additional Bottle Cap?" Part 2: Discussion of Kaitlyn's Work

[A few seconds of turn-and-talk]

Ms. Gordon: What did you talk about with your partner? I was talking to Ellis and Gabriel who had some interesting observations, and I was listening to other groups. Let's share with the class: What did you notice [unintelligible] . . . Ellis?

Ellis: She put numbers in the circles. She had 8 in the circle. She put 8 in circles. Also she put 12 circles. She added 1 more and that equaled 13. It equaled 21.

Ms. Gordon: Wait, wait. You said "equals 13" and then you said "equals 21."

Ellis: She put 1 more, that it had 13.

Ms. Gordon: What do you mean, now that had 13?

Ellis: I mean she put those numbers [pointing], also that other number [unintelligible].

Ms. Gordon: OK. Can anyone say more about what Ellis is saying?

Josiah: That she added 1 more.

Ms. Gordon: OK, she added 1 more. How can you tell?

Josiah: Because there's 1 more down there.

Ms. Gordon: I heard Ryan—you were saying something's confusing.

Ryan: [pointing to the circled 1 underneath the row of 8 circles]: Oh, because this is 1. There's two 1s here.

Ms. Gordon: What do you mean? Say more.

Ryan: Because I said it was confusing because there's two 1s there. I was like, was that supposed to be 9?

Ms. Gordon:	OK, why might you think that might be 9?
Ryan:	I thought that was supposed to be 9.
Ms. Gordon:	OK, why did you think that was supposed to be 9?
Ryan:	Because, like, two 1s—there's no two 1s in the numbers.
Ms. Gordon:	OK, so you see an additional 1 over here. There's two 1s over here. You're thinking that this 1 should be 9? [Ryan nods.] OK, so Tunmishe, can you add on to what Ryan's saying?
Tunmishe:	She added 1 to the 8 instead of 1 to the 12.
Ms. Gordon:	So from this representation, it's making you think that the additional bottle cap is with this group of 8? So you're thinking this should be, like, 9?
Josiah:	It should be like over there, and it should say 13.
Ms. Gordon:	OK. Why?
Josiah:	Because, like, it's kind of confusing. Cause it looks like she like started over, like she's counting back from 1.
Anu:	It looks like she added 1 more. If the circle is [moved] and she puts 13, everybody would think you added 13 [unintelligible].
Ms. Gordon:	Added 13 to what?
Anu:	To the group.
Ms. Gordon:	So I'm hearing a few different ideas. We can tell that an additional bottle cap was added, from our initial problem, Ms. Gordon had 8 bottle caps, Ms. Schooler had 12. But some friends are saying [pointing to the circled 1 under the group of 8 circles] it makes it seem like the additional bottle cap was added to maybe this group here, Ms. Gordon's. And some are thinking it should have been over here to make it clear that the additional bottle cap was added to Ms. Schooler's. Huh.

9.3. "Where Is Ms. Schooler's Additional Bottle Cap?"
Part 3: Discussion of an Equation

As the discussion continues, Ms. Gordon asks the class to compare the equations in the two representations. Livia's equation is $8 + 12 + 1 = 21$. Kaitlyn's equation is $20 + 1 = 21$. We rejoin the class a couple of minutes later into the discussion.

Tierra:	It doesn't look the same, but it does have the same answer.
Ms. Gordon:	OK, so the total is the same. One of them had the total of 21. This one has the total of 21.

Anu:	Kaitlyn's equation has the full total of the first equation, but she just added 1 more to make it 21.
Ms. Gordon:	So we knew from the first problem that their total . . . they had 20 bottle caps altogether. And then with the second part of the problem when Ms. Schooler found another bottle cap, she just added 1 to the total, and now there's 21 altogether. Well, then I have a question. What if either one of them wrote this? [Writes $8 + 12 + 1 = 20 + 1$.] Can you do that? Does that make some sense?
Students [chorus]:	Yes.
Ms. Gordon:	Why? I heard the "yes." J'aimeson?
J'aimeson:	Because I know that the 20, and she added 1 more, so it can make 21.
Ms. Gordon:	Well, what about this whole, the whole equation? So you said 20 plus 1 equals 21, but my equation is $8 + 12 + 1$ equals $20 + 1$. Does this whole thing make sense? [Some students say yes.] Why?
J'aimeson:	Equations are like that.
Ms. Gordon:	OK, can someone say more?
Tunmishe:	That 8 plus 12 plus 1 equals 21, and that they're just like a different number. It's just with a different number.
Joniah:	So we had 8 plus 12 equals 1, but I recognize it equals 21 because this is 12, 20, and it added one more to go to 21.
Anu:	Yes, because this sign, it doesn't only mean equals, it means "the same as," and 20 plus 1 equals 22, and . . .
Ms. Gordon:	Wait, 20 plus 1 equals?
Anu:	I mean 21, and 8 plus 12 plus 1 equals 21, too.
Students:	Now I agree . . . I was going to say that.
Ms. Gordon:	21 is the same as 21. In Livia's representation, she wrote the equation $8 + 12 + 1 = 21$, and Kaitlyn came up with the same total. She used our initial total of 20, she added 1 more for the one more bottle cap that was found, and now there are 21 bottle caps. So both equations came up with, had the same total, and if I put them together, both expressions together to make one equation, they equal the same thing. 21 is the same as 21. OK? Awesome. So very nice job to everyone. I just picked two scholars' work to share, so thank you, Livia and Kaitlyn, for allowing us to talk about your work. And very nice job to everyone for being engaged and participating in our discussion.

Chapter 10

Facilitating Critique and Revision of Student-Created Representations

"First draft" is a familiar term in writing. After a first draft is written, the author reads over their story, essay, or poem to correct spelling and grammatical errors, considers whether a description is apt, and perhaps reorders sentences or paragraphs to restructure the flow of ideas. A poetry teacher of one of the authors once said, "You know what makes a good poem? Thirty or forty drafts!" And this process is rarely performed alone. In almost any book, you can find in the acknowledgments section a list of friends or colleagues the author thanks for insights, suggestions, or corrections made on early versions of the manuscript. With such feedback, authors revise their work in order to better communicate with readers, and it often takes listening to a range of perspectives to get it right.

"First draft" also applies to mathematics. The idea of revising drafts in mathematics may seem strange to those who view the content as cut and dried, right or wrong. In that view, revision might only mean that students should look over their work to see if they made a mistake in a procedure or wrote a number incorrectly. But when we think of mathematics as the exploration of ideas, the notion of revision becomes more complex. For example, in Chapter 7, we saw students working together, revising classmates' wording of a conjecture to make sure it communicates what they intend. Similarly, students might revise other forms of representation—drawings or story problems, for example—to capture the ideas they want to convey.

> **In this chapter, which focuses on revising drawings and story problems, you will**
>
> - analyze a set of representations to consider which need revision to more clearly reflect a conjecture about multiplication,
> - view a video of a fourth-grade class discussion in which students revise some of those same representations,
> - consider commentary offered by Critical Friends, and
> - read the teacher's writing in which she explains her decision to include representations that require revision among those she presents to the class.

As you do these activities, we'd like you to think about these three aspects of the principle that *student-created representations offer anchors, openings, and depth.*

1. As with conjecturing, an individual's drawing, diagram, or story problem becomes a tool for collective activity with the purpose of better conveying mathematical relationships.

2. Students come to understand that in mathematics it's not only appropriate, but also necessary, to offer partially formed ideas on which to build.

3. To encourage the development of strong mathematical ideas and a disposition toward digging deeper, the teacher selects pieces of work that both offer important math ideas and require revision.

Do the Math

Creating a drawing, diagram, or story problem to illustrate a conjecture is more complex than creating a representation to illustrate a single calculation. When illustrating a conjecture, one must demonstrate how the representation of one equation is related to and can be transformed into the representation of the other. For example, recall Livia's representation from Chapter 9, which illustrates the conjecture, *If I add 1 to an addend, the sum goes up by 1.*

Figure 10.1 • Livia's Representation of 8 + 12 = 20 and 8 + 13 = 21

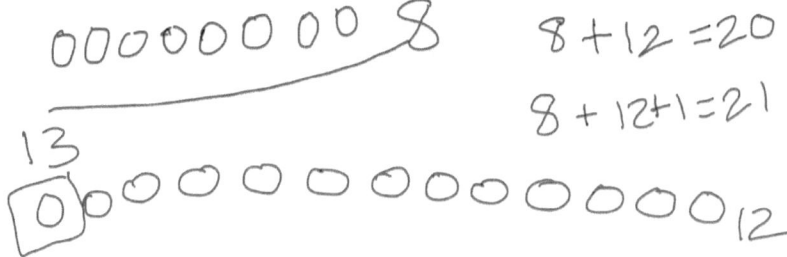

1. How many do they have together?

First, the drawing shows the relationship among the parts of the equation, 8 + 12 = 20: 8 circles and 12 circles can be joined to make 20 circles Then, Livia drew an additional circle to show the increase of 12 to 13, creating a representation of 8 + 13 = 21. The parts of the equation, 8 + 12 = 20, are embedded within the drawing of 8 + 13 = 21. By putting a square around the circle labeled "13," Livia shows 12 circles within the collection of 13, and how increasing an addend by 1 necessarily also increases the sum (the two quantities joined) by 1.

We're now going to turn to a group of fourth graders working on multiplication. Like the third graders in Mr. Fairley-Pittman's class in Chapter 7, they are working on what happens to the product when a factor is increased by 1. They have been creating representations to illustrate their class conjecture: *When I add a 1 to the first or second factor, the product goes up by the factor that does not change.* Michelle Sirois, their teacher, asked students to show how the conjecture works using the following pairs of equations.

$3 \times 5 = 15$	$3 \times 5 = 15$
$4 \times 5 = 20$	$3 \times 6 = 18$

She then created the poster shown in Figure 10.2 on which she reproduced the student representations she selected to share with the class. (Note that in the row of four squares with the numeral 5 in each square, the first three squares were drawn with one color marker, and the fourth square was drawn with a different color. Similarly, the bottom row in the array of squares on the left and the last column in the array of squares on the right were drawn in different colors from the rest of those arrays.)

Figure 10.2 • Ms. Sirois Redrew Selected Representations Created by Students

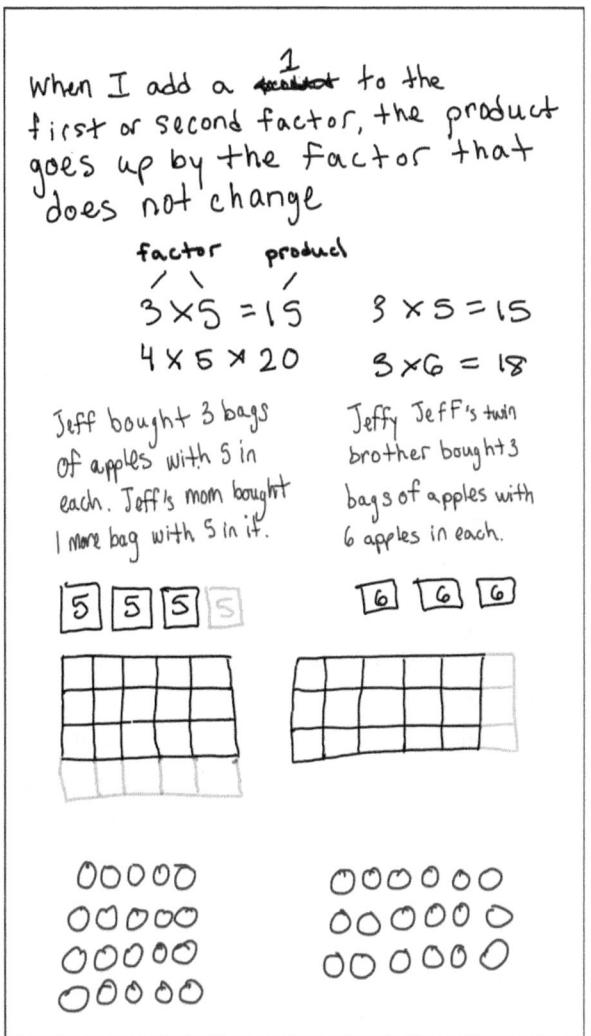

For each of the drawings and story problems in the poster, consider these questions:

1. How does the representation show multiplication?

2. Where do you see a factor increase by 1?

3. Where do you see the increase in the product?

4. How does the representation show the conjecture?

5. Does this representation need revision in order to illustrate the conjecture more clearly? If so, how would you revise the representation?

Watch the Video: "Could I Change My Writing?"

At the start of the lesson, Ms. Sirois covered up the representations illustrating the second pair of equations, asking the class to focus on the representations of $3 \times 5 = 15$ and $4 \times 5 = 20$. For each representation, students addressed the core questions:

- How does the representation show multiplication?
- Where do you see one factor increase by 1?
- Where do you see the increase in the product?
- How does the representation show the conjecture?

The class decided that, in the array of circles (bottom left of Figure 10.2), a box should be drawn around the bottom row in order to show how the 3×5 array is embedded in the 4×5 array, thus more clearly illustrating the conjecture. Ms. Sirois next removed the cover over the second set of representations to discuss the representations of $3 \times 5 = 15$ and $3 \times 6 = 18$. The class was satisfied with the clarity of the array of squares (the first 5 columns were drawn in blue, the last column in green).

The video consists of three segments. We first hear Ms. Sirois asking whether some revision is needed to the representation of three boxes, each containing "6," and to the story problem. After a turn-and-talk, which you won't see on the video, the next segment shows students revising the box representation. Finally, in the third segment, one student, Abdoulaye, revises the story problem. (Note that Abdoulaye had created the representations of boxes and the story problems in both columns on the poster.)

First Viewing of the Video: "Could I Change My Writing?"

Watch all three segments of Video 10.1 with a focus on what students are learning.

Video 10.1

"Could I Change My Writing?"

qrs.ly/hofs4yr

Reflecting on the Video:
The Mathematics Students Are Working On

[You may want to use the transcript at the end of this chapter as you consider these questions.]

1. How does the students' revision of the drawing of three boxes containing the numeral 6 more clearly illustrate the conjecture?

2. How does the revision of the story problem more clearly illustrate the conjecture?

3. How does this lesson deepen students' understanding of multiplication?

4. How does this lesson deepen students' understanding of the conjecture and the act of representing a conjecture?

Second Viewing of the Video:
"Could I Change My Writing?"

Now rewatch the three video segments with a focus on how an invitation to revise might affect students' sense of agency.

Reflecting on the Video:
The Role of Revision

[You may want to use the transcript at the end of this chapter as you consider these questions.]

1. What moves did Ms. Sirois make to invite students into the practice of revising?

2. What do you notice about Abdoulaye's participation in the revision of his representations?

3. When students revise a classmate's work, how might that process affect the classroom community?

Read and Reflect on
What Others See in the Video

Let's return to the three aspects of the principle that *student-created representations offer anchors, openings, and depth* listed at the beginning of this chapter.

---

1. As with conjecturing, an individual's drawing, diagram, or story problem becomes a tool for collective activity with the purpose of better conveying mathematical relationships.

2. Students come to understand that in mathematics it's not only appropriate, but also necessary, to offer partially formed ideas on which to build.

3. To encourage the development of strong mathematical ideas and a disposition toward digging deeper, the teacher selects pieces of work that both offer important math ideas and require revision.

When our Critical Friends viewed the video, their comments centered around two main themes: the value of presenting multiple representations and how Ms. Sirois's moves shifted in different parts of the discussion.

1. What Are Students Learning About Multiplication?

In Chapter 7, when Mr. Fairley-Pittman's class was working on a conjecture related to the distributive property, our Critical Friends noted the depth of the mathematical content. In this lesson, in which a class was analyzing representations of a similar conjecture, the Critical Friends commented about how the mathematical ideas went still deeper.

Yi Law Chan: In what ways are these students uncovering deep mathematics and constructing ideas together? Why did Ms. Sirois take time to connect the story to the notation to the array and to the boxes that show groups—all the visuals on the poster? What do the students gain?

Lynne Godfrey: I remember once when Deborah and Susan Jo were leading a session at one of our study groups and asked us to write a story for the conjecture about adding 1 to a factor. I remember thinking, can I make one story to show the change, or do I need two different stories? That was a big idea for me. And how could I draw one representation to show both equations? That really did challenge me. Having the equations, the story, and the array all in alignment was exciting for me to see.

Virginia Bastable: Looking at the equations and even the conjecture, it might seem easy, and you think you understand it, but you don't really until you can do this other step, where you can show all the different components of the conjecture in the story and the other representations. Lynne is saying there was something more to figure out about something that she already knew at one level. To me that clarifies what Yi Law is getting at when she asked about "deep mathematics."

Hetal Patel: These story contexts are different from the way we usually think about story problems. It's not that you're supposed to come up with an answer. Here, they're using stories to represent an idea, just like the arrays or the pairs of equations or the pictures of groups. So the story is another representation of the idea. It was worthwhile for the class to spend time on each of the representations, to see how the same idea can be represented in varied formats.

Darlene Ratliff: There's an equity aspect to this, too. If students are presented with *only one* form of representation, what about those who don't immediately see it that way? With all the different forms of representation, students have opportunities to find their way into the mathematics as they understand it. And then, having the mathematical discussion, looking across all the representations to see where to find the same components in different representations, and revising those that need revision—all of this provides opportunities for all students to move deeper into the mathematical ideas.

2. Teacher Moves and Student Participation

In this part of the conversation, the Critical Friends discussed the moves Ms. Sirois made to invite students into close examination and revision of the representations, and they considered how the moves a teacher might make depend on the specific circumstances of that moment.

Virginia Bastable: At the beginning of the clip, Ms. Sirois calls on Maeve to say how she would revise the drawing of three boxes, and then she calls on Sarah to write out Maeve's idea. It felt like a lot of people were engaged in changing the representation with the boxes.

Hetal Patel: I highlighted that part, too. When Sarah went up to the board, the teacher asked, "What do you think Sarah could do? Watch her." She was inviting other students to get inside Sarah's head and participate mentally. And when Sarah asked if she could cross something out, Ms. Sirois said, "If you want." The teacher's word choice gave the agency back to Sarah.

Cindy Ballenger: Ms. Sirois trusted the students to mark up the poster. I was wondering, if I were the teacher, what would I do if they didn't make the change in the direction I was hoping?

Virginia Bastable: But when Abdoulaye was changing his story problem, Ms. Sirois was in control of the marker. I noted how carefully she wrote exactly what he said, and then she invited him to edit.

Cindy Ballenger: At first Abdoulaye was not very explicit, and then he became more explicit. Was she careful to write down his words because she had confidence that it would get clearer or because she wasn't sure where he was going? Maybe she had to stick with exactly what he said in case he went off in a different direction.

Yi Law Chan: I was curious about how Ms. Sirois responded to Vivian's question about Abdoulaye's change in the story problem. She could have asked Abdoulaye to clarify or she could have asked another student for a thought. We have seen in other occasions where she assigns agency to students, but here she stepped in. I thought it was interesting that she jumped in saying, "I think I did a disservice here." I wonder about the impact of the teacher taking ownership of a possible "disservice" on the learning community.

Marta Garcia: When Vivian said she was confused, you can see in Abdoulaye's body language that he was losing some confidence in the work he had done. Then Ms. Sirois pretty quickly lifts up his idea for the group. I wondered if she wanted to be sure that his idea was clear because he had taken a risk.

Ms. Sirois seems to be paying attention to nonverbal cues throughout the clip. Teachers who understand that verbal communication is just one way to communicate ideas are tuned in to signals that support them in making on-the-spot decisions. This awareness of multiple forms of language supports all learners in gaining access to and articulating their ideas.

Virginia Bastable: And then Ms. Sirois asked this wonderful question afterward, inviting Abdoulaye to explain what made him want to change his story.

Lynne Godfrey: I thought that was a nice way of giving agency back to Abdoulaye. I love the words he used: "The representation was not going the same way as my story."

Reflection Questions

1. Yi Law Chan raises these questions: "Why did Ms. Sirois take time to connect the story to the notation to the array and to the boxes that show groups—all the visuals on the poster? What do the students gain?" How would you respond?

2. Several Critical Friends comment on how Ms. Sirois is writing down what students say or making changes they are suggesting. How do you think about recording students' ideas? What if those ideas aren't, as Cindy Ballenger suggests, going in the direction you were hoping for? Are there times when you might modify what a student says when you are recording? For what reasons?

3. Several Critical Friends pondered Ms. Sirois's moves after Vivian said she was confused about Abdoulaye's revision of his story context. List at least two possible moves Ms. Sirois could have made at that moment. For each move, consider the following questions:

   • Under what circumstances might that have been a good move?

   • How would that move have supported students' learning?

Teacher Reflection

Ms. Sirois later wrote a profile of Abdoulaye, giving us more background on her goals for him and her choice to present his work to the class, knowing that it would need revision.

Michelle Sirois: Abdoulaye is seen as a leader by his classmates. He is very capable and has good ideas. He likes to completely re-do assignments when he realizes something isn't quite right. Although in some ways getting everything right can be a strength, I also want Abdoulaye to come to recognize the value of offering incomplete ideas and to develop his ability to use feedback as an opportunity for learning. I want him to look for more than confirmation that he has an answer right or wrong. For Abdoulaye, the lesson sequences that focus on generalizations about the operations have been an opportunity for him to understand that learning can be about experimenting and revising.

In the lesson in which students created representations showing how 3×5 is related to 3×6, Abdoulaye had drawn three boxes with the numeral "6" in each box, and he had written a story problem about Jeffy having three bags with six apples in each. Neither of these representations explicitly showed 3×5. Several other students had created representations similar to Abdoulaye's, all needing revision. When I was selecting samples of student work to share in the discussion shown in the video, I felt it was risky to choose Abdoulaye's representations, but I decided it was worth the risk.

The next day during our whole-group discussion, classmates started to revise the representations. For Abdoulaye's representation of three boxes, they collectively chose to show three groups of 5 + 1 rather than three groups of 6. During the next turn-and-talk, Abdoulaye's voice had a sound of defeat and disappointment, and I encouraged him to consider and learn from the revisions his classmates suggested. Abdoulaye stayed quiet for the next part of the conversation, and I thought I would be having another check-in with him later around building his capacity to use feedback for learning. However, a few minutes later, Abdoulaye stated that he wanted to revise the story to match the representation. With that, he changed it to read, "Jeff bought 3 bags of apples with 5 in each. Jeffy, Jeff's twin brother, bought 3 apples and put them in Jeff's 3 bags that have apples." This moment was a turning point for Abdoulaye in his ability to see feedback from others as an opportunity to learn and revision as an opportunity to make an idea clearer.

In the coming sessions, I noticed Abdoulaye participated frequently, confidently speaking up, interjecting with a thought on his mind. In another lesson, he and his partner started the conversation by asking questions of a classmate's representation, unclear of what it represented. Throughout the conversation, they worked to make sense of what was being represented and suggested adding labels to the representation to make it clearer. At one point when a classmate was confused, Abdoulaye asked, "What do you not understand?," trying to make sense of his classmate's confusion and offer an idea.

I've noticed as the year went on, the process of revision became easier for students. I'm wondering whether, when revising together, perhaps they don't see it as "I'm wrong" but rather "We're making these representations better as a group."

 **Reflecting on Ms. Sirois's writing**

1. Ms. Sirois wrote that when Abdoulaye revised his story problem, "This moment was a turning point for Abdoulaye." How do you understand what shifted for him?

2. What was the nature of the classroom community that allowed that shift to happen?

3. What opportunities for learning might that shift have opened up for him?

What Do You Want to Remember From This Chapter?

Take a few minutes to note for yourself ideas you want to hold onto as you continue to investigate the meaning of a mathematics community and how to build it. What teacher moves have you noticed in this chapter that you want to bring into your own practice? Here are some of the ways we like to think about what the Collaborating Teachers and Critical Friends in this chapter have demonstrated to create openings for every student:

- **Encourage students' expression of incomplete ideas.** Students' incomplete ideas might be key to their and other students' understanding of a significant concept.

- **Engage the class in the process of revision.** Among the goals for teaching is helping students learn to receive critical feedback and to participate with their classmates in revisions of their own work.

Taking a Next Step

Consider the students' representations you collected at the end of the last chapter or collect a set of student work now in which you present a story problem to students and ask them to create a drawing, diagram, or arrangement of physical objects. If you were to lead a discussion based on three samples of student work, which work would you select? In making the selection, consider the dual commitment to equitable participation and significant mathematics. Explain your reasoning for each selection.

Video 10.1 Transcript: "Could I Change My Writing?"

Ms. Sirois:	So here's what I want to talk a little bit about, though, because some of you were confused. I want to go back to Callie's thing. I hear you talking about . . . you're talking and using this array a lot to connect to the conjecture, what's changing. And I'm wondering do we need to make some changes to these representations to actually show the change, because right now this shows 3 × 6, but I'm a little confused where the 3 × 5 is. And right now this shows 3 × 6, but I'm a little confused where the 3 × 5 is. So what could we maybe change in our representations or adjust to make them clear that we're showing that change. Talk with your partners about that.
	[Students talk with their partners. The video continues after the turn-and-talk.]
Ms. Sirois:	I was talking to Maeve and Kaylanis and they have an idea for how they might change this representation to show both the 3 × 5 and the change that is happening to 3 × 6, so Maeve or Kaylanis, does one of you want to start and we can add on from there. What were you suggesting?
Maeve:	We were saying that like [inaudible] the 3 fives and put 1 in each because that would be showing . . .
Ms. Sirois:	What is Maeve talking about? Does somebody want to come and try and change what she's talking about? Sara, do you want to come and try and change the representation?
Sara [at chart paper]:	Wait, can I like x this out and change [inaudible]?
Ms. Sirois:	Yeah, if you want. [Sara xs out each 6 and writes a small blue 5 in each box.]
Sara:	And then maybe with a different marker . . .
Ms. Sirois [while Sara gets a different marker]:	What do you think Sara could do? What could Sara do now? Watch her. What is she doing? [Sara writes a small green "+1" in each box.]
Some students:	Ooohhh.
Ms. Sirois:	Fatima, what is she doing?
Fatima:	So, she's adding, so she took away the 5, she took away the 6 . . .
Ms. Sirois:	Talk to your classmates, not to me . . .
Fatima:	She took away the 5 from the 6 in each box and then there's going to be 3 more so she put the 1s in each box. [Fatima

	goes up to the chart paper and continues as she points.] She crossed out the 6 and she put a 5. There's 3 more apples so she added 1 to each bag.
Vivienne:	So I want to add on to Fatima because it's talking only about Jeffy right now, but so right now we're trying to show the 5×3 and the 6×3 so you need to be able to show the 5 and then adding 1 to show what Jeffy has.
	[After some further discussion, we rejoin the class as Abdoulaye raises his hand.]
Abdoulaye:	Could I change my writing?
Ms. Sirois:	Do you want to change your story now? [Abdoulaye nods] How would you change your story?
Abdoulaye:	Um, Jeff's twin brother, Jeffy, bought 3 apples and added them with the other bags.
Ms. Sirois:	And added them with what other bags?
Abdoulaye:	The bags that Jeff and his mom bought, that Jeff bought.
Ms. Sirois:	So keeping this the same, so would you say this again over here . . . so [correcting the story on the chart] Jeff bought 3 bags of apples with 5 in each. Jeffy, Jeff's twin brother, bought 3 apples . . . and what?
Abdoulaye:	3 apples and put them in the bags.
Ms. Sirois:	And put them in . . .
Student:	in the 3 bags.
Ms. Sirois:	the 3 bags?
Abdoulaye:	the 3 bags that have apples.
Vivienne:	I'm a little bit confused about that because if he's only getting 3 apples, he doesn't have those 3 bags anymore, so he can't add them to the 3 bags.
Ms. Sirois:	This was set up yesterday. Abdoulaye was starting with this story [pointing to the first story about Jeff and the 3 bags] for both situations, and I probably did a disservice by not putting this part here. So if we read it like this, "Jeff bought 3 bags of apples with 5 in each. Jeffy, Jeff's twin brother, bought 3 apples and put them in the 3 bags that have apples." So whose 3 bags is he putting them in?
Students:	Jeff's
Ms. Sirois:	In Jeff's 3 bags [she edits chart so that it reads "put them in Jeff's 3 bags"]. Does that make more sense if we put these two parts together to show? [Students: yes] Abdoulaye, I appreciate you being willing to make a change. What made you make the decision that you wanted to change your story?

Abdoulaye: Because I didn't feel like my representation was going the same way as my story, and it was confusing.

Ms. Sirois: You didn't feel like this representation was matching the story as much? Do we feel like this now represents our conjecture? Does this show the change in our factor now? And are we showing the change that happens to our product? [Students nodding and saying "yes"] We have had a very long "Why Does It Work?" conversation today, and we know our next step, and we're always thinking, does this always work? What numbers will it work for? And these representations and stories that we're talking about today, could we use them with other numbers and other scenarios? We're not going to talk about that now. We're going to take a wrap and move around for a little bit.

Part Four

Students Are Initiators and Advocates for Their Own Learning

Think back to some of the students you have encountered in this book: In Chapter 3, Jeff Parks described how his student, Makayla, created on her own a single diagram to illustrate the relationship between multiplication and division. Toward the end of the video clip of that same class, you saw Aleeyah recognize how the multiplication/division relationship is analogous to the addition/subtraction relationship. Or consider Guled in Chapter 5, who contrasted two different ways of changing a subtraction equation: "If we only added or subtracted to one of the numbers in the [subtraction] equation, then there would be a different [difference]. But if we added to both the numbers in the equation, then it would still be the same." These students are going beyond the questions or tasks their teachers set out for the class, taking on their own challenges, posing their own questions, making connections to earlier learning, and taking initiative to extend their thinking.

When a classroom community is built on the expectation that students engage in mathematical reasoning and have mathematical ideas, when students are invited to express their thoughts and build on the thinking of their classmates, students will continue to have more ideas, building a stronger and stronger web of connections that may go beyond the teacher's goals for a particular lesson. They demonstrate their mathematical agency by asking such questions as "What if I change this in the problem?" "What happens if I use a different kind of number?" "What can I see in a different representation?"

Throughout the book, there are other examples of students demonstrating agency. We hear students who say, "Please move so I can see," and "I didn't hear what she just said," knowing that it matters. And there are those who speak up when they don't follow someone else's point. Once students know what it feels like to make sense of mathematics and have had experiences of moving from confusion to understanding, they engage the content with the expectation that they *will* understand. They come to recognize that confusion is not an end state or a signal to stop trying. Rather, it is a step in the process of figuring things out: Acknowledging and considering confusion can open up the space for new insights. In the words of Duckworth (1996), "All of us need time for our confusion if we are to build the breadth and depth that give significance to our knowledge" (p. 82). If expressing confusion is recognized as a contribution to the mathematics community, students will ask the class to help them make sense.

In Part Four, we extend our understanding of mathematical agency as we focus on students who take initiative to set questions and challenges for themselves and become advocates for their own learning. We consider implications of students taking such agency for the teacher and for the classroom community.

The following are the major themes of Part Four:

- In a mathematics community that focuses on student thinking, students become initiators and advocates for their own learning.

- When students introduce new ideas or ask for help to clarify their own understanding, they contribute to the learning of the mathematics community as a whole.

- Teachers can set up norms and structures to affirm the variety of students' ideas that might be generated during a lesson while maintaining the coherence of the lesson for all students.

Chapter 11

Supporting Students to Take Charge of Their Own Learning

A third-grade class was discussing this set of equations:

$12 + 8 = 20$	$12 + 8 = 20$
$14 + 8 = 22$	$12 + 10 = 22$
$38 + 45 = 83$	$38 + 45 = 83$
$43 + 45 = 88$	$38 + 50 = 88$

They had already formulated a conjecture for what happens to the sum when one addend increases by 1, and they had created pictures, story contexts, and arrangements of manipulatives to explain why their conjecture was true. Now they were extending their generalization to consider what happens when *any number* is added to an addend. Some students had noticed that, given $12 + 8$, when 2 was added to either addend, the sum increased by 2, and given $38 + 45$, when 5 was added to either addend, the sum increased by 5. As different students were working to find language to describe their observations, one student, Danielle, declared, "I just noticed something! When you add two even numbers you get an even number, and when you add an even and an odd number, you get an odd number."

At this moment, the teacher may have been facing a dilemma. On the one hand, pursuing Danielle's idea right then and there would have meant taking on a different math idea, perhaps cutting off the participation of classmates who were thinking about adding different amounts to an addend. On the other hand, moving forward with the planned agenda may have suppressed Danielle's excitement and exploration of her conjecture about odd and even numbers.

However, in this case, there was no dilemma. In this classroom in which students were encouraged to approach mathematics with curiosity, the teacher had set norms for managing the variety of ideas students presented. "What an interesting thought," she responded to Danielle. "Write it down on your scratch pad." Everyone in the class knew what she meant. They each had a "scratch pad," a place where they could record an idea when the class wasn't going to pursue it at that time. Sometimes students would write it down for themselves to continue thinking on their own, and sometimes they'd show their scratch pad to the teacher, who kept a poster on the wall that listed ideas that were still open questions for students to ponder. From time to time, students would wander over to the poster to read the ideas that had been offered. The teacher used the poster to stay alert for opportunities to return to some of those ideas.

In this chapter, we consider classrooms that cultivate students, like Danielle and her classmates, as initiators of their own learning—students who pose their own questions and set challenging problems for themselves.

> **In this chapter, you will**
>
> - do some mathematics and pose questions that might extend your own learning,
> - view video from a fourth-grade lesson in which pairs of students are working on their own, and
> - consider comments about the video from our Critical Friends and the teacher of the lesson.

As you consider the classroom video and the commentaries in this chapter, we'd like you to think about three aspects of the principle that *students become initiators and advocates for their own learning*. These three aspects describe how all students are encouraged to participate and to dig deeply into their own learning of mathematics.

1. Each student is learning to find the edge of what they know, pose their own questions, find their own challenges, and move their own thinking forward.

2. Such an orientation on the part of students is the product of the community's interest in students' insights, questions, and developing skills.

3. Classroom norms and structures can be put into place to manage the variety of thoughts and issues that may arise as students express their own ideas.

Do the Math

Prepare for viewing the video of a fourth-grade lesson by investigating the generalization the class is embarking on.

Consider the following pairs of equations:

$6 \times 8 = 48$	$14 \times 10 = 140$	$18 \times 50 = 900$
$12 \times 4 = 48$	$28 \times 5 = 140$	$9 \times 100 = 900$

1. What do you notice about these pairs of equations?
2. Write a conjecture that is suggested by these pairs of equations.
3. From the work you did in #1 and #2, what are next questions you might pursue?
4. Share your conjecture and next questions with a partner.

Watch the Video: "I'm Thinking of Ideas"

In a previous lesson, Michelle Sirois gave to her class the six multiplication problems in your math activity without the answers. Once students had solved the problems on their own, she created a poster with the six problems and asked a pair of students to fill in the answers. At the start of this lesson, Ms. Sirois told the class they were beginning a new sequence of noticing patterns and articulating conjectures. "So say you were given these equations," Ms. Sirois said, "what do you notice about these pairs of problems?"

After a turn-and-talk, students offered their ideas. Some students talked about *multiples*—"12 is a multiple of 6 and 8 is a multiple of 4"—a term that had recently entered their vocabulary. Sophie said, "6 + 6 = 12 and 4 + 4 = 8." And Sara said, "6 when it's doubled is 12, and 8 when you cut it in half is 4, and that gives you the same product." As the discussion continued, different students worked to put their thoughts into words, offering different versions of either what Sophie was indicating—thinking of the relationships additively—or what Sara said—thinking of the relationships multiplicatively.

A big shift in thinking for third- and fourth-grade students involves noticing and working with multiplicative relationships. They are more used to thinking, for example, that 15 is *3 more* than 12 than that 15 is *3 times more* than 5. That was a major issue at play in the discussion of what Ms. Sirois's students noticed. Ms. Sirois was hoping that their work on this new pattern would help more students to recognize that they could *multiply* one factor by 2 and *divide* the other factor by 2.

In the video, we enter the class after about 20 minutes of discussion when Ms. Sirois sets up a task for students to work on in pairs. You will see Ms. Sirois explain the task, and then you will get glimpses of several pairs of students working on the task: Sara and Fatima, Abdoulaye and his partner, and Sophie and Conal. The clip ends with whole-group discussion. During the pair work, because students are working close together, you won't be able to hear everything that's said. From what you can hear and what you can see on their white boards, pay attention to the variety of math ideas the students bring up, as well as what is indicated about their engagement with the work.

First Viewing of the Video: "I'm Thinking of Ideas"

Watch Video 11.1 with a focus on the mathematics students are working on.

Video 11.1

"I'm Thinking of Ideas"

qrs.ly/mnfs4yz

Reflecting on the Video: The Mathematics Students Are Working On

[You may want to use the transcript at the end of this chapter as you consider these questions.]

1. What do you notice about the numbers students choose to test out the pattern they've identified?

2. What mathematical questions arise for students about the patterns they've noticed?

3. What do you make of Abdoulaye's comment, "I'm thinking of ideas"? What does it indicate about his sense of what math class is about?

4. The task provides an opportunity for assessment. What can Ms. Sirois learn by listening to pairs, looking at students' white boards, and hearing what they offer after their work in pairs?

Second Viewing of the Video: "I'm Thinking of Ideas"

Now rewatch the video clip, keeping in mind the question of what Ms. Sirois might do with all the ideas and issues that arise from this task.

Reflecting on the Video:
Managing Many Ideas

[You may want to use the transcript at the end of this chapter as you consider these questions.]

1. If you were in Ms. Sirois's position at the end of this lesson, what might you do next? Why? (Think back to the vignette that opens this chapter. What are some strategies you might use to honor the ideas even if you can't deal with them all in the near future?)

2. What opportunities do you see here to support and extend students' deep mathematical learning?

3. What opportunities do you see here to support and extend equitable participation?

4. What opportunities do you see here to support and extend the development of students' mathematical agency?

Read and Reflect on What Others See in the Video

Let's return to the three aspects of the principle that *students become initiators and advocates for their own learning*:

1. Each student is learning to find the edge of what they know, pose their own questions, find their own challenges, and move their own thinking forward.

2. Such an orientation on the part of students is the product of the community's interest in students' insights, questions, and developing skills.

3. Classroom norms and structures can be put into place to manage the variety of thoughts and issues that may arise as students express their own ideas.

When our Critical Friends viewed "I'm Thinking of Ideas," their comments centered around two themes: (1) the mathematics content and practices students were working on and (2) the challenges students' initiative presents for the teacher.

1. Mathematical Content and Mathematical Practices

While viewing the video, our Critical Friends were noting both the specific concepts and skills students were working on, as well as how students' agency as mathematicians is visible in their work.

Marta Garcia: Fatima used the word "addition" to describe how the factors are changing. She's still making that critical shift to start thinking about multiplicative relationships. But I also want to say that this clip shows how this work not only illuminates the structure of the operations, but it's also an opportunity to increase computational fluency.

Lynne Godfrey: There's so much stuff happening in the video. Like some students were thinking 3 × 11 doesn't work because you can't divide 3 by 2. That's a good question for fourth graders to think about. They saw the pattern as long as they stayed in the realm of whole numbers. Then a classmate reminded them that you can move into decimal numbers: 3 ÷ 2 = 1.5. And then there were the students who were choosing numbers that were larger than they knew how to work with. What happens when you multiply multiples of 10 or multiples of 100? How many zeroes are in your product?

Yi Law Chan: Even while we see there's more for these fourth graders to learn about place value, they're now working on how to make equivalent multiplication problems. They're now extending the concepts they explored in addition and subtraction to multiplication, and that's deep and rigorous mathematics. The mistakes in place value we saw them making—that's still part of the fourth-grade standards. They'll continue to work on that throughout the year. For now, there was so much understanding in the way students were trying different factor pairs to make equivalent problems. And then they were asking "what if" questions like, What if the factors are the same? What happens when you try 30 × 30?

Virginia Bastable: It felt like the students had the mindset of what it means to do an experiment. When I watched Abdoulaye become intrigued by his question—what happens if the factors are the same?—I was reminded of my daughter when she discovered commutativity of multiplication. She ran around saying, "6 × 2 is the same as 2 × 6," and "3 × 8 is the same as 8 × 3," and then one afternoon she came out of her room laughing, saying, "Listen to this! 6 × 6 is the same as 6 × 6!" She thought it was very strange because when she switched the positions of the 6s, it looked just the same. I thought Abdoulaye might have been intrigued by something similar. Abdoulaye didn't finish the problem. Once he got the calculation right, he would see that 30 × 30 = 900 and 60 × 15 = 900. What's important here is that he was curious and asking the question.

Hetal Patel: We saw Abdoulaye on video in an earlier chapter and read Ms. Sirois's profile of him. Back then she said Abdoulaye was concerned about having right answers, and she wanted him to learn to try out ideas and revise his thinking. Now we see him selecting problems that are stretching his learning and feeling almost gleeful about experimenting and having ideas.

Lynne Godfrey: We see in this video how the students have taken on that orientation to their learning. Formulating their own questions and challenges grows out of the community's interest. As they take on each other's questions, they'll work together to address them.

2. The Challenge to Teachers of Managing a Classroom That Promotes Students' Individual Agency

Our Critical Friends recognized that teachers who are new to focusing on students' thinking are likely to be uncomfortable with the possibility that so many ideas arise at once and that the lesson will have to end with questions left up in the air. As they discussed this, they thought about their own experiences as classroom teachers as well as their work to support teachers' professional development.

Lynne Godfrey: I have always believed that if you leave it open to the kids, they will choose just the right place to start from. When I saw this video, I thought, oh my goodness, look at all the stuff that's happening. There's a lot for the teacher to learn about her students from what they're doing and what they choose to focus on. But I could see someone else might see the video as order breaking into chaos when you let your students have a choice of what to do.

Virginia Bastable: I can see it might be really scary for teachers who feel their job is to make sure there's closure on this.

Lynne Godfrey: When I was teaching, there were times I'd give the class a challenging problem, and they were trying things all over the place, but almost all of those places had important mathematics in them. And so each place was something to come back to—not all at once, not even by tomorrow.

Teacher Reflection

Ms. Sirois, herself, did not see chaos in this lesson. In fact, she was impressed by the thinking of her students, what it showed about their comfort as mathematical thinkers, and how the notion of lingering on ideas had become a feature of the community. She began her reflection on how her students engaged more generally with the lesson sequences on generalizations about the operations.

Ms. Sirois: When I look back on the video of these lessons, I see how the lesson sequences on generalizing had normalized unfinished thinking for our community. Students know that we're going to continue thinking about this. We're going to try other numbers, and we're going to write a conjecture, and we're going to create a variety of representations to show how the conjecture works.

I often keep the anchor charts up for a while. I like students to see how we worked on a conjecture, with all the revisions on the poster, and to have all the representations and story problems. Students would sometimes reference them in later lessons.

At some point, after the suggestion from a coach, I started creating a poster for our lingering questions. We'd just leave it up, and if a new idea came up, we'd add it. Sometimes we'd return to those questions in a later conversation. But also, students could go to that poster and do some thinking about lingering questions. We were keeping those questions alive.

When I looked at the videos again, I saw kids being comfortable playing around with numbers. In the lesson about equivalent multiplication expressions, they wanted to think about big numbers, numbers in the thousands. And they wondered about odd numbers, which brings them into fractions and decimals. Doing this work gets them comfortable exploring and playing with numbers, which is an important skill to have.

This work also allows you to differentiate. At times, a student might begin to ask, "How can I generalize this?" before the rest of the class was ready to take on that question, and I would encourage them to explore that on their own.

Students came to value all the different ideas their classmates were bringing up. Toward the end of one lesson, someone said, "We should do a gallery walk." They were suggesting that student pairs create posters to show all the different numbers they tried and then include representations and story problems for them. They were already thinking about the next steps we take when we explore a generalization.

I've come to see that there are mathematical goals around the concepts, but there are also goals about how students are as mathematicians. A while back, I wrote about Abdoulaye. He was a student who could grasp mathematical concepts quickly, but he still needed to learn to be okay being flexible, okay with not being right, okay with revising, okay with experimenting. When I viewed the video about equivalent multiplication expressions, I especially tuned into how he was as a mathematician, how willing he was to experiment. That's a goal I hold for all of my students.

Reflecting on Building Students' Agency

Ms. Sirois wrote, "[T]here are mathematical goals around the concepts, but there are also goals about how students are as mathematicians." What goals do you hold for your students as doers of mathematics?

What Do You Want to Remember From This Chapter?

Take a few minutes to note for yourself ideas you want to hold onto as you continue to investigate the meaning of a mathematics community and how to build it. What teacher moves have you noticed in this chapter that you want to bring into your own practice? Here are some of the ways we like to think about how teachers support their students as initiators of their own learning:

- **Expect a variety of ideas.** When classrooms are focused on student thinking, students will offer more and more ideas that cannot be addressed all at once.

- **Encourage lingering questions.** Some questions are worthy of time to ponder and don't require closure within a single lesson.

- **Affirm student agency.** Students may, at times, offer ideas that seem to be "off topic," but the having of ideas is to be encouraged *for itself*. Establish norms and structures that provide ways to affirm students' ideas while still enabling the flow and coherence of the mathematics students are learning.

Taking a Next Step

During one of our discussions with our Critical Friends, Yi Law Chan asked, "How does a teacher build students' sense of responsibility to their own and their classmates' learning, the way we see in the videos and Ms. Sirois's comments?" As you think back on the chapters of this book, the video you've seen, the teachers' reflections, and your own notes and reflections, what are your responses to Yi Law Chan's question?

Video 11.1 Transcript: **"I'm Thinking of Ideas"**

[Note: Because students are working close together in pairs, transcription of students' words was more difficult than in most of the videos. We've given you as much as we could, but we think the ideas different students are working on will be clear.]

Ms. Sirois: I'm going to push you right now. You and your partner, you're starting to notice this pattern. All of these pairs have something in common, and you're starting to articulate what's happening if you multiply one factor by 2 and divide another factor by 2, what's happening with the product? I want to see if you and your partners can come up with more pairs of equations that do this. Okay? So see, you and your partner are going to work together. I'm going to give you a marker board and a marker. [Student: What partners? Ms. Sirois: Your rug partner.] And you're going to see . . . can you come up with more pairs of equations that work like this? You can work with small numbers. Maybe you want to try with some larger numbers. You're going to explore right now. Okay?

Sara and Fatima have 9 × 4 = 36 on their whiteboard and have started the next equation with 18 ×.

Fatima: It's 2.

Sara: It can't be 2 because when we, um . . .

Fatima: [unintelligible] 18 times 2 equals 36.

Sara: Oh! [Sara records "× 2 =," then hesitates]

Sara: We can't do that. We can't use that.

Ms. Sirois: Does this follow the same pattern? Go back to the pattern [unintelligible].

Fatima: This one's adding, this one's minusing.

Sara: No, well yeah, but this one's multiplying by 2, this one's dividing by 2.

Ms. Sirois: What did Sara say?

[Fatima starts restating, then Sara begins to clarify—unintelligible]

Figure 11.1 • Sara and Fatima's Whiteboard at the End of Class

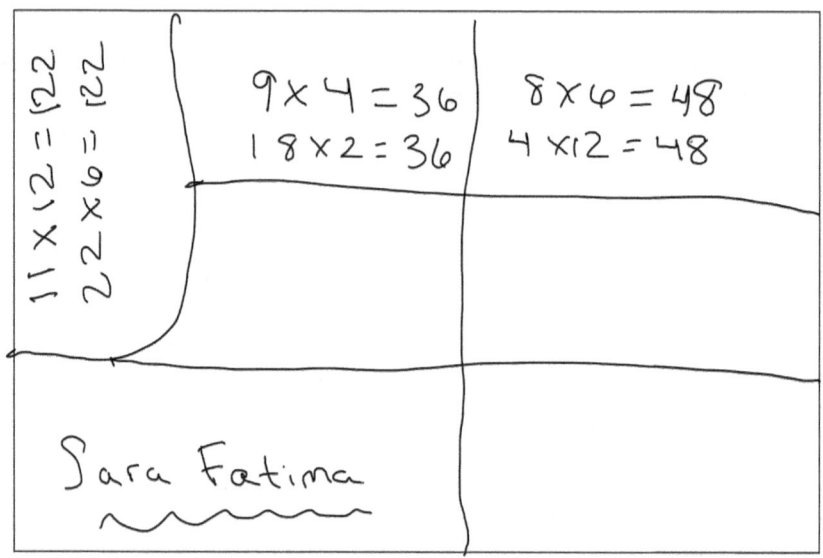

$11 \times 12 = 122$
$22 \times 6 = 122$

$9 \times 4 = 36$
$18 \times 2 = 36$

$8 \times 6 = 48$
$4 \times 12 = 48$

Sara Fatima

Abdoulaye: Ms. Sirois! [He shows his board, then continues working.]

Abdoulaye: I'm thinking of ideas. I think you can do two of the same number. Uh, two times. I'm going to try. [He writes 30 × 30 = 90 and continues writing, while his partner makes suggestions.]

Figure 11.2 • Abdoulaye's and His Partner's Whiteboard at the End of Class

$$55 \times 100 = 550$$
$$x^2 (110 \times 50)^{÷2} = 550$$

$$12 \times 6 = 72$$
$$x^2 (24 \times 3)^{÷2} = 72$$

$$30 \times 30 = 90$$
$$x^2 (60 \times)^{÷2}$$

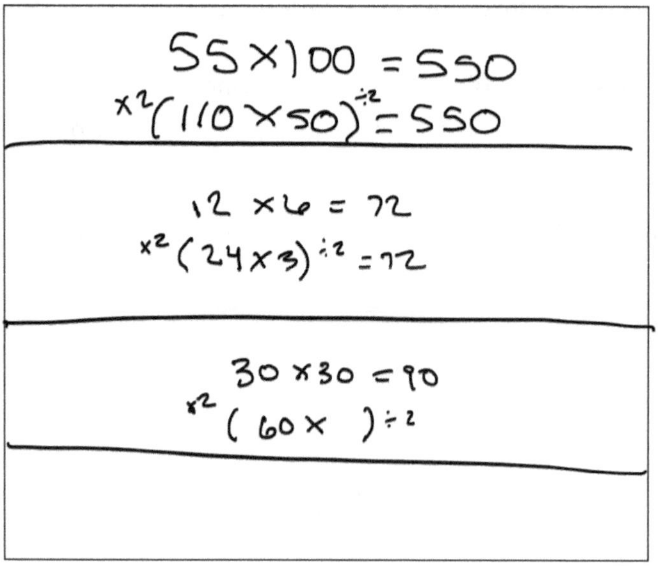

Sophie and Conal have a pair of equations on their whiteboard: 60 × 500 = 3000 and 30 × 100 = 3000. They are considering whether to change the 100 to 1000.

Ms. Sirois: Did you start with 1000? Why are you changing it?

Conal: Well, actually I did have it as 1000.

Ms. Sirois: You did?

Conal: Yeah.

Ms. Sirois: OK, can you fix it? What needs to be fixed to 1000, Conal?

[Conal directs Sophie to change 100 to 1000 on their board:] Put a zero there.

Ms. Sirois: Why are you making that change, Conal?

[Conal explains, pointing to the numbers [(unintelligible)].

Ms. Sirois: So something's happening to the factors. Can you explain what's happening to the factors?

Sophie: So he's doing half of 60 is 30, and half of 1000 is 500.

Figure 11.3 • Sophie and Conal's Whiteboard at the End of Class

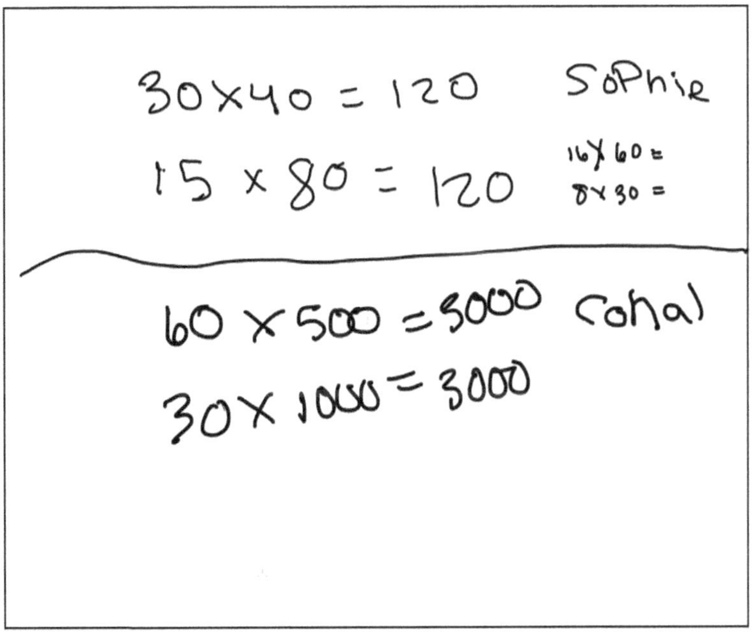

$$30 \times 40 = 120 \quad \text{Sophie}$$
$$15 \times 80 = 120 \quad \begin{array}{l} 16 \times 60 = \\ 8 \times 30 = \end{array}$$

$$60 \times 500 = 3000 \quad \text{Conal}$$
$$30 \times 1000 = 3000$$

. . .

Ms. Sirois: I want to see what everyone wrote on their board because our next session we're going to think about, Did you find pairs of equations that work? We're also going to think about, Did you get to some numbers that you felt weren't working for you? Did anyone feel like that—they got to some numbers that weren't working or some factors that weren't working? Fatima?

Fatima:	So me and Sara were going to do 11 times 3, but we knew we couldn't, so we wanted to split the 3, but then we were, like, we couldn't do that.
Conal:	You could've did 1.5.
Fatima:	Yeah, could be minus.
Ms. Sirois:	Interesting. What did you say, Conal?
Conal:	You could do 1.5.
Ms. Sirois:	Huh. Interesting, Conal, you think, oh, we just learned yesterday how we read that. Fourth grade, do you remember?
Conal:	I wasn't here yesterday, so . . .
Ms. Sirois:	Oh, you weren't here yesterday. Darn it—you got me! We'll talk about this tomorrow. This we read as one and five tenths or one and a half. You can read it two different ways. Interesting. I'm going to leave this here for us to think about and grapple with if Conal's idea is going to work for us. Abdoulaye?
Abdoulaye:	I was trying an idea. I was experimenting, that what if you did two of the same numbers?
Ms. Sirois:	Can you give me an example?
Abdoulaye:	Like 30 times 30?
Ms. Sirois:	What would happen in that case?
Abdoulaye:	Um, I tried it.
Ms. Sirois:	Did you try it? With your experiments?
Abdoulaye:	Yeah, well, I didn't finish it.
Ms. Sirois:	Okay, I'll leave that there for us also to continue exploring.
	[some remarks, unintelligible]
Ms. Sirois:	One last thought before we wrap up for today.
Sophie:	Um, me and Conal both, Conal was writing something and he, instead of writing, . . .
Ms. Sirois:	Wait one second, we're still having a conversation over here. Why don't you come sit here, it'll help you. Thank you.
Sophie:	So we did 60, I mean we did 60 times 500, um, and then we did 30 times 100, but then he had to make the change and turn it into 1000. That equaled 3000.
A student (Conal?):	I know it's not the answer.
Ms. Sirois:	Alright, here is what we need. We need to pick up—don't erase your boards, please. Please, please don't. Just leave them nicely there. I'll give, I won't yell, I will, um, we'll make sure we take a picture of your work.

Chapter 12

Normalizing Confusion

A second-grade class had been investigating the relationship between addition and subtraction. On this day, the teacher presented to the class the poster in Figure 12.1, which shows the story problem students had been given in an earlier lesson and one student's drawing to represent it.

Figure 12.1 • A Student's Drawing to Represent the Story Problem

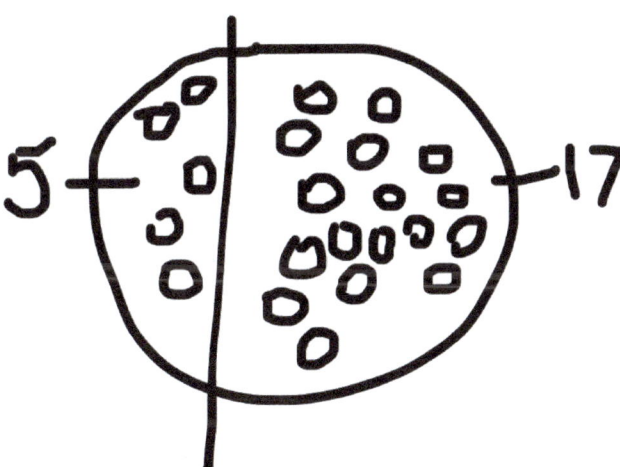

> Meg is holding 22 balloons. 17 fly away. How many balloons is she holding now?
>
> 22 – 17 =

The teacher opened the discussion with the question, "What do people think about Ethan's representation?" The class understood they were not being asked about the answer to the story problem; they all knew the answer was 5. Nor were they being asked to describe their own drawing. Rather, the question was about how to interpret Ethan's drawing as a representation of the problem.

Ryan made the observation, "I'm thinking that 17 plus 5 equals 22. In Ethan's drawing, there's 22 singles in there, and then there's minus the 17, and then there's still 5 in the 22, so that's how many he got left with."

Other students didn't yet see how Ethan's drawing represented the problem; nor could they follow Ryan's point. But as the discussion continued, first one girl called out, "Oh, I get it. I totally get it!" and in the next several minutes, ahs of understanding rippled through the class. Only Briana said she didn't see it. When invited to say what she was confused about, she said she expected 22 on one side of the circle, 17 on the other side of the circle, and then a separate group of 5.

Students continued to describe what they saw in the representation and explained how it showed not only $22 - 17 = 5$, but also $5 + 17 = 22$. The teacher periodically checked in with Briana to see if anything was clicking for her, and students offered suggestions for what might help Briana understand, but Briana was still shaking her head. By now Ethan's diagram no longer simply belonged to Ethan. It was a tool for the whole community as they suggested additional labels to illustrate the relationships more clearly.

When it was time to end the discussion, the teacher said to Briana, "Sometimes with something like this, you have to walk away from it, and it cooks in your brain when you're not paying attention to it, and the next time you come to look at it, you see it a little bit differently." The teacher was speaking from her own experience learning mathematics. Over the last several years, as she had been digging more deeply into mathematical concepts, sometimes she, too, had to walk away from a problem, and when she came back to it, she would have new insights. She had also learned that understanding wasn't always like a light switching on and staying on. She described how, if a new concept wasn't yet solidly integrated, she would "go in and out of understanding."

From experiences as a math learner herself, the teacher knew not to be dismayed when her students didn't understand. In fact, she encouraged them to say when something didn't make sense. She wanted her students to come to see that confusion is a step toward understanding, and expression of confusion is an opportunity for the class to linger on a set of ideas that are worth the time. As different students explain what they see, understanding deepens for everyone—even if clarity isn't achieved in a single lesson.

In this chapter, you will

- reflect on moments of your own confusion as a mathematics learner,
- view video from a fourth-grade class in which a student expresses confusion, and
- consider comments by our Critical Friends and the teacher in the video.

As you engage with the material in this chapter, we'd like you to think about three aspects of the principle that *students become initiators and advocates for their own learning*. These three aspects of building community lay a strong foundation for interweaving rigorous mathematics and equitable participation.

1. Confusion is a normal and frequent phase of learning.

2. Expression of confusion is assumed to contribute to the class's understanding.

3. Mathematics belongs to the community, not to individuals, and everyone is responsible for each other's understanding.

Do the Math

Before viewing the video, take a moment to reflect on your experiences as a math learner by addressing the following:

1. Review the work you've done on the math activities in this book, and identify a moment that you were confused about the mathematics. If you don't think you were ever confused while working on this content, think back over your entire experience as a math learner to identify a moment of confusion.

2. What did you find confusing? If you are no longer confused, what helped you sort it out?

3. Share your moments of confusion with a partner. Discuss what kinds of moves from a teacher or classmates are helpful or not helpful.

Watch the Video: Fatima's Question

The video you are about to see was taken a few minutes after the video viewed in Chapter 10, when Michelle Sirois's fourth-grade classroom was discussing the representations shown in this anchor chart.

Figure 12.2 • The Anchor Chart Ms. Sirois Presented to the Class at the Beginning of the Lesson

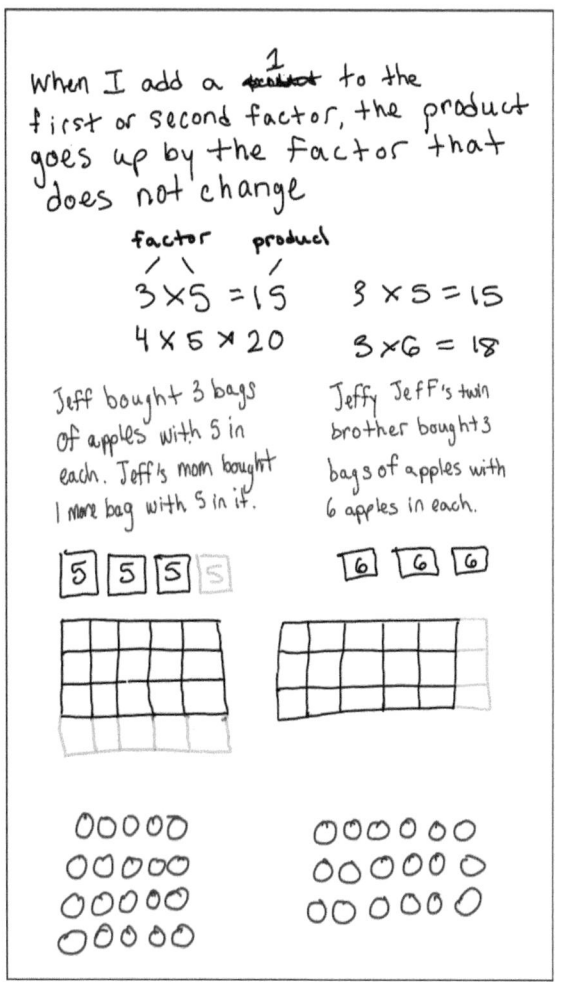

To review what happened in that lesson, the class focused on the three boxes with the numeral 6 in each. They decided that, in order to illustrate the conjecture about adding 1 to a factor, the 6s should each be replaced by 5 + 1. On the basis of that discussion, Abdoulaye, who had written the story problems, changed the story about Jeffy. His new version read, "Jeff bought 3 bags of apples with 5 in each. Jeffy, Jeff's twin brother, bought 3 apples and put them in Jeff's 3 bags that have apples." At this point, Ms. Sirois decided to wrap up the discussion so that students could move around for a bit before they needed to get to their next activity.

In the video we are about to see, which happens just after Ms. Sirois has tried to end the lesson, we discover that one student, Fatima, wasn't ready to let the discussion end and was eagerly raising her hand. We'll rejoin the class as Fatima asks her question: "Why would Sara cross out the 6 if she's going to put 5 and plus the 1? Because that's 6, so what's the point?"

First Viewing of the Video: Fatima's Question

Before viewing Fatima's Question, you may choose to review Video 10.1 seen in Chapter 10, which is found here:

Video 10.1

"Could I Change My Writing?"

qrs.ly/hofs4yr

Watch Video 12.1, "Fatima's Question," with a focus on the mathematics content students are working on.

Video 12.1

Fatima's Question

qrs.ly/97fs4z1

Reflecting on the Video: The Mathematics Students Are Working On

[You may want to use the transcript at the end of this chapter as you consider these questions.]

1. Following a 30-minute discussion, some of which was seen in the clips in Chapter 10, Fatima opened up the discussion again, and in the next three minutes, five students spoke. In terms of learning mathematics, what was the value of these three minutes to individual students and to the class as a whole?

2. What additional information about student understanding might the teacher have learned from this part of the discussion?

Second Viewing of the Video: "Fatima's Question"

Now rewatch the video clip with a focus on the nature of the learning community in this classroom.

Reflecting on the Video:
The Nature of the Learning Community

[You may want to use the transcript at the end of this chapter as you consider these questions.]

1. How did Fatima advocate for her own learning?

2. How did students respond to Fatima's expression of confusion?

3. What does the students' discussion imply about what it means in this class to do mathematics together?

4. In Ms. Sirois's comments at the end of the video, what messages was she trying to communicate to her students?

Read and Reflect on
What Others See in the Video

Let's return to the three aspects of the principle that *students are initiators and advocates for their own learning* listed at the beginning of this chapter.

1. Confusion is a normal and frequent phase of learning.

2. Expression of confusion is assumed to contribute to the class's understanding.

3. Mathematics belongs to the community, not to individuals, and everyone is responsible for each other's understanding.

When our Critical Friends viewed Fatima's Question, they first commented on the nature of the classroom community as illustrated by its response to Fatima. Then they focused on the new and sometimes surprising information revealed by watching this clip after having seen the clip from Chapter 10.

1. The Nature of the Classoom Community

 **Yi Law Chan:** Fatima's question—"What's the point?"—could have been construed very differently in the context of another kind of environment. Here, the response was, wait, there's something more to understand, and her classmates really took that on. In this class, they know they spend time making connections and looking carefully at representations.

 **Lynne Godfrey:** Nobody in this class made a disparaging gesture, even though it had been a long conversation, and the class was supposed to go somewhere else.

 **Virginia Bastable:** You could see the teacher looking at her watch, like she knew someone else was expecting them, but she also knew this discussion was important.

 **Marta Garcia:** This clip illustrates the culture of the classroom taking a stance of curiosity toward the mathematics. They know that asking questions is what mathematicians do, and they don't have to be satisfied with incomplete understanding. Members of their community can advocate for themselves.

 **Hetal Patel:** In response to Fatima's question, multiple kids came up and continued the conversation without the teacher chiming in.

Virginia Bastable: Ms. Sirois usually doesn't tell students when they're right or wrong. But when Fatima was speaking too softly for her classmates to hear, Ms. Sirois asked her to speak up and assured her that what she was saying made sense. If Fatima didn't want to speak loudly enough because she wasn't sure of herself, Ms. Sirois removed that obstacle.

2. It's Not Over 'Til It's Over

After viewing the clip included in Chapter 10, our Critical Friends had made several assumptions about the understanding of some individual students and of the class as a whole. They found that these assumptions were challenged when they watched the discussion that ensued after Fatima asked her question. A community in which students take the initiative to ask questions and express confusion continually illuminates students' understanding for the teacher.

Virginia Bastable: I'm just thinking, it's not over 'til it's over. We were all following that conversation in the Chapter 10 clip, and a lot of students were explaining to each other and working hard. As a teacher, I could easily have felt satisfied with where the class was. And then one student was strong enough to keep saying, "Why did you do that?" It's an amazing reminder of the complexity of this work when you're dealing with a lot of individuals all working on the same idea.

Lynne Godfrey: When Abdoulaye first changed his story problem in the earlier clip, he still seemed shaky about the changes in his story. Now, a few minutes later, he was so crystal clear, showing what in the story problem connected to what in the drawing of boxes, and explaining why it's necessary to change the 6 to 5 + 1.

Cindy Ballenger: In the earlier clip from this lesson, we saw Fatima explain why Sara changed the 6 to 5 + 1, but a few minutes later, something happened that made her say, wait a minute, why did Sara do that? But that happens to all of us, right? We answer a question and then stop and ask, How do I know that?

Hetal Patel: Yes, we all have that experience of going in and out of understanding.

 **Reflecting on Confusion as a Phase in the Process of Understanding**

1. If your students seem disparaging or impatient with another student's confusion, how might you handle that, both in the moment and over time?

2. After seeing the video and reading comments by our Critical Friends, return to what you discussed in the math activity. What further thoughts do you have about helpful responses on the part of a teacher and classmates when a student expresses confusion?

3. Teacher Reflection

When reflecting on these three-and-a-half minutes from her lesson, Ms. Sirois lets us in on her observations, motivations, and new insights, all grounded in knowledge about the individuals in her class.

Ms. Sirois: This clip left me with many lingering thoughts. One important aspect is how multiple students responded to Fatima's question. As a facilitator of the conversation, I tried to stay out of it with my own reactions and ideas. Even though Sara explained her thinking, and I thought Fatima may have understood at that point, I allowed three more students to share their reaction to Fatima's question. Doing so had multiple functions: (1) I was able to assess how other students were making sense of Fatima's question and the representations. (2) Just one person's explanation may not have made sense to Fatima or other students. By having multiple students speak, other students could hear the argument in many ways, one of which may have made more sense than another. 3) Multiple students had the opportunity to verbalize their understanding and see themselves as important members of the mathematics community. One student who spoke in this series is new to our classroom community, having just joined the class a couple of weeks ago, and has been designated to get special support in math. The fact that she strongly stated what she understood is just one example of how these conversations can build community and boost students' mathematical identities.

This clip also highlighted the importance of using stories as a form of representation. When Abdoulaye responded to Fatima's question, he referred back to his story to explain his reasoning. The story allowed him to see the change in the factors and product, and after he explained I could hear at least one student go "ohhh. . . . " At multiple other moments throughout the conversation, students referred back to the story to explain how the visual representations were showing the conjecture.

The use of students' language and advocacy in this clip were interesting to me. First, the conversation continued because Fatima advocated for her question to be heard and discussed. At one point in the clip, another student voices that she can't see when Abdoulaye is explaining. It may seem minor, but in this moment, a student who infrequently participates in the conversation was making known that she was listening and trying to make sense, too. Also, when watching the clip back and transcribing what was being said, I noticed how students were saying "so *we* can show" and "*we* wanted to show" and "*you guys* wanted to show." Even though it started as one student's representation and one student's revision, the class was taking collective ownership over the work. To me, this use of language demonstrated that students saw the process of revision as theirs to do together.

Reflecting on Collective Ownership of the Mathematics

Ms. Sirois noted that the class was taking collective ownership of their work. Go through the transcript, and identify words and actions that provide evidence of her observation.

What Do You Want to Remember From This Chapter?

Take a few minutes to note for yourself ideas you want to hold onto as you continue to investigate the meaning of a mathematics community and how to build it. What teacher moves have you noticed in this chapter that you want to bring into your own practice? Here are some of the ways we like to think about what the teacher and Critical Friends in this chapter noted about supporting students' advocacy for their own learning of mathematics:

- **Normalize confusion.** Refrain from avoiding confusion or "saving" students when they seem to be confused. Confusion is often a step toward understanding. Rather, invite students to say more about what they are confused about, or take them back to ideas that feel solid for them. At other times, let students know that they may simply need time away from a problem and will have new insights when they look at it again.

- **Normalize lingering questions.** Just as we often need time to sort out confusion, some questions require time to ponder.

- **Affirm student advocacy.** Students express advocacy for their learning in many ways: when they share their thinking, when they pose questions, when they say they are confused. Sometimes advocacy is more subtle, such as when they ask a classmate to speak up or to move so they can see.

Taking a Next Step

Review your notes from working with this book, and think about the themes and commentaries that have especially struck you. What issues of practice stand out? Have you made any changes in your own practice in light of your reflections? If so, what are they?

Video 12.1 Transcript: **Fatima's Question**

Fatima: Why would Sara cross out the 6 if she's going to put 5 and plus the 1? Because that's 6, so what's the point?

Ms. Sirois: Sara, do you want to explain?

Sara: Because it shows both equations. Because it changes the color, so the "plus 1" is a different color, so it shows 5×3, and then when you add, and it changes the color to plus 1, and then it's 6×3. So it can show both equations.

Michael: It's showing the change by $5 + 1$ into 6.

Abdoulaye: Can I come up? I wanted to say it's because since Jeff had 3 bags of apples with 5 in it, it still keeps the 5 with the 3 bags. And then Jeffy, he bought 3 apples, so then he put them in each bag, so that's why she crossed out the 6 so then we can show that 5, that we had 5 before and Jeffy put the 3 apples in.

Samihya: I understand what Sara did right here when she crossed out the 6, when we showed 3×6. She crossed out the 6 and put 5 so it could show 3×5 and 1 more so it could show 3×6.

Fatima: So at first it was only showing 6×3 but you guys wanted to show 5×3, so you did, you crossed out the 6 and then it shows both, because if you plus those together [points to the 5 and the 1], it's that [points to the equation 3×6 written on the chart paper], and if you don't, if you don't really care about the 1, it's going to be 5×3.

Ms. Sirois: You've got to speak up louder. You're making sense, but your classmates can't hear you, and you're saying to them, you're saying hey, you are helping me understand this. So say it loudly. Say what you understand now.

Fatima: So you guys said that you guys wanted to show both equations, 3×5 and 3×6. And now I get how you guys are showing it because at first when Sara didn't cross out the 6, it was only 6×3, but we wanted to show the 5×3, too. So Sara crossed out the 6, and put a 5 and then a 1, so it shows both because if you plus those together it's the 6×3 and if you don't plus the 1 to the 5, it's 3×5.

Ms. Sirois: Is that what you guys were trying to do? [Students: "yes"] You helped Fatima make sense. I love it. A lingering question that we had answered. Fatima, way to be bold and brave and still ask it. Alright, we'll wrap it up there for today, for right now.

Conclusion

What To Do Next

Which images from the classroom stay with you as you think back on this book? Perhaps Ryan comes to mind, the first grader in Chapter 9 who explains to the class why he can't make sense of Kaitlyn's drawing. Are you remembering third grader Hailey in Chapter 6 who builds on her classmates' ideas to clearly state a conjecture about subtraction or fourth grader Fatima in Chapter 12, who insists on asking her question? Or did you notice students who never spoke or seemed, in those moments of the video clip, disconnected from the classroom discourse?

In centering unscripted video in this book, our intention has been to provoke reflection on classrooms as they are in real time, with the teachers making decisions in the moment, and with the variety of student behaviors evident in any classroom—the students who look engaged and those who don't, the students who are wiggly and those who aren't, the students who appear to be articulate and confident and those who have difficulty finding words. In our extensive use of classroom video examples over several decades, we've found that viewers can be quick to assume they are seeing some kind of "special" classroom, sometimes insisting, "My kids could never do that." The six untracked, public-school classrooms represented here add to a growing body of images of the strong mathematical capabilities of students who have been historically underserved and underestimated in mathematics, including girls, Black and brown students, multilingual learners, and many other students who have been assumed to be incapable of mathematical reasoning. In these videos, we can see the moments of brilliance and insight, the strengthening of mathematical identity, and the enactment of agency as math learners, all of which occur regularly in classrooms where the teachers are focused on equitable participation in challenging mathematics.

Our purpose is not to stand back to admire or critique what is happening in these classrooms but rather to offer these examples so that other educators can reflect on what goes into building equitable mathematics communities in their own contexts. Our six Collaborating Teachers would be the first to say their classrooms were far from perfect—that, in fact, there is no "perfect" but only continued questioning, experimenting, reflecting, and taking new actions.

As you read this book and viewed the videos, we hope that you felt that you were in conversation with us, with our Collaborating Teachers, and with our Critical Friends—all of whom brought their own perspectives to the interweaving of equitable participation and rigorous mathematics content. Based on the main ideas of the four parts of the book, there are four foundational principles that can guide your continued thinking:

- A mathematics community that is focused on deep mathematics and equitable participation allows all students to develop both individual and collective agency as mathematics learners.

- Deep and complex mathematics requires collaborative construction of ideas. Discussion that is focused on deep mathematics involves a nexus of ideas, offering students different entry points and different modes of participation.

- Representations in the form of diagrams, pictures, arrangements of physical objects, and story contexts support deep understanding of the mathematics. Student-created representations allow students who have different ways of thinking about the mathematics to express and explain their ideas.

- In a mathematics community that focuses on student thinking, students become initiators and advocates for their own learning.

While we suggested next steps at the end of each chapter, this book is not a step-by-step guide for what to do, but it is intended to provide a basis for ongoing dialogue, questioning, challenge, and reflection. In this concluding chapter, we don't offer additional goals or take-aways, but we suggest strategies for your own lifelong learning that we have seen make a difference in teachers' practice.

When we came to write this book, we were determined that it must weave together a profound respect for children's mathematical thinking, rigorous mathematics content, and a focus on nurturing the voices and ideas of every student. We knew we could not do this alone—the collaboration with the six teachers is what made this book possible. But what made *their* classrooms possible? What is it about these six teachers that led to their determination to both teach deep mathematics content and provide access to that content for every student?

As we have worked with many teachers across several decades and in many different contexts, we have seen how teachers with these goals focus their own ongoing study in three areas: making sense of mathematics for themselves, nurturing their curiosity about and understanding of student thinking, and honing their attention to each student's participation.

Making Sense of Mathematics for Yourself

Teachers need to learn—and continue to learn—mathematics. Making sense can't be a goal for students if teachers have not worked on making sense of the content for themselves. For us, a book on mathematics community that focuses on students making sense of challenging math ideas had to include the opportunity for the adults using the book to dig into the math content for themselves. That's why most chapters include a "Do the Math" section, and we hope you took the time to engage with the problems in those sections—not only in order to have a preview of the math students were doing in the video you were about to see but also because engaging in what might appear to be "easy" mathematics *always*, in our experience, opens up deeper concepts and connections for adults' math learning. Further, these sections are an opportunity to simply enjoy doing mathematics.

Many of the teachers we've worked with over many years initially learned mathematics as a set of facts, definitions, and procedures to be remembered: Success in mathematics meant getting the correct answers to assigned problems, often the faster the better. Even today, this view of mathematics persists. Based on such a view of the nature of mathematics, many Grades K-5 teachers, and especially women, the majority of elementary teachers, decide— reflecting what is common in our society in general—they aren't "math people." But given the opportunity to work on mathematics in an environment that emphasizes making sense of mathematics, teachers begin to see mathematics as an interwoven network of ideas to be explored. They recognize that they have mathematical ideas, that those ideas develop further as they, perhaps working with colleagues, examine different avenues to solving a problem, and that they can make connections between concepts they had previously considered isolated facts. In projects in which teachers had this opportunity, they discovered that doing mathematics can be deeply satisfying, and as they exercised their own new-found mathematical agency, they became thirsty to learn more. For many teachers, these experiences opened a new world of possibilities for themselves as mathematics learners and as effective teachers. As described in the Preface, the six Collaborating Teachers at the heart of this book received preparation and support through the Boston Teacher Residency (BTR). Part of BTR's curriculum was a strong mathematics component, focused on learning mathematics content, learning about students' mathematical thinking, and emphasizing sense-making in the mathematics classroom.

How might you continue to learn mathematics? Perhaps you can enroll in an online course or a program offered within your school system or a local college or university. You might be interested in the professional development series *Developing Mathematical Ideas* (see Appendix B), which we produced precisely for teachers to dig deeply into the mathematics content they are responsible for teaching. Perhaps you can form a study group to work on problem sets with colleagues. Sometimes teachers meet with colleagues to investigate the mathematics in the curricular materials they teach from. For example, Ms. Sirois talked about learning from the lesson sequences on noticing and conjecturing she was using in the videos you've viewed. The sequence about what happens to the product when you add 1 to a factor goes on to explore what happens when you add not just 1 but any whole number to a factor, and then what happens when you add a fraction to one of the factors. Ms. Sirois commented, "This work really allowed me to see the connection between whole numbers and fractions and how we can use the same representations for both."

Nurturing Curiosity About and Understanding Student Thinking

In Chapter 1, we began with two ideas:

1) Mathematics is an interwoven network of ideas.
2) Students come to school with mathematical ideas, and part of the work of the teacher is to draw out those ideas and help students develop them further.

A key factor driving us in the creation of this resource was our own curiosity about students' ideas—how students express themselves, what they say, what they draw, what they see in others' drawings. We believe that one factor that leads to equitable mathematics learning is *the teacher's curiosity* about how students understand the mathematics and how students perceive themselves as learners, both as individuals and as part of a larger math community. The realization that students arrive in school with their own mathematical ideas provides an exciting opening to investigate students' thinking.

Teachers need opportunities to nurture and express their curiosity about student thinking. Learning about student thinking largely happens through careful observation of one's own students, examining their written work and listening intently to their explanations. With a stance of curiosity, a teacher's questions communicate respect to students. Learning about student thinking also deepens our own understanding of the mathematics, as when we are trying to understand a student's novel approach to a problem. For example, one teacher reported how her second graders were working on the problem $37 - 18$. When one student, Brian, started by saying, "30 take away 10 is 20," she assumed he was in trouble, that his next step would be $7 - 8$, which can't be solved with whole numbers. But Brian surprised her when he went on to explain, "7 take away 8 is 1 in the hole; 20 and 1 in the hole is 19." The teacher said that she had since learned that this is one of several common methods second graders use to subtract, but it still felt new enough to her that she had to think it through each time. "This is definitely a case," she said, "of learning mathematics from my students."

But how does a teacher even remember what students have said by the end of a busy day that has comprised hundreds of interactions? In order to increase knowledge of student thinking, teachers need to somehow slow down the process of that thinking in order to reflect on it. In many of the professional development projects we have led, including the project on which this book is based, we've encouraged teachers to audio or video record classroom dialogue or interviews with individual students, and to write "cases" of classroom lessons based on the recordings. Teachers report that reflecting on one short segment in the flow of a day's interactions allows them to dig deeply, to learn more about their students, and to consider implications for their teaching. Having to write out their students' words, they discover things were said they hadn't heard in the moment and realize that if they continue to listen hard, some of their assumptions about their students will be challenged.

At the same time, recording students' words and work communicates curiosity about students' ideas. Once students realize how carefully their teacher is listening, they offer more ideas, precisely because they are given the opportunity to express them. They often ask about what teachers are writing and are proud that their ideas are being recorded. It's not uncommon for students to say something like, "Don't you want to turn on your recorder? I'm about to say something important."

Sharing student work, interviews with students, or classroom cases with colleagues on a regular basis provides the opportunity for learning about student thinking and how students develop identities as doers of mathematics.

Honing Attention to Each Student's Participation

Throughout this book, our purpose has been to illuminate examples in which teachers are focused on raising up their students' voices and ideas, thinking through each student's needs and strengths, and planning for ways to bring each student into the classroom community.

In contrast, some years ago, in a classroom one of us was observing, four students of color sat together at a table in the back of the room, a room that was longer from front to back than it was wide. When it was time for a whole-class discussion about the math work, the teacher stood at the front of the room and called mostly on white students sitting toward the front, never calling on any of the four at the back table, despite eagerly raised hands. It is easy to stand outside of this scenario and condemn the teacher's actions. But we also need to consider: Had this teacher ever been asked to notice who was participating in the classroom, whom they called on and didn't, and what kinds of questions were directed at different students? Had the teacher ever been challenged and supported to reflect on this aspect of their practice?

These questions lead us to ask whether we have ourselves been that teacher at some time, to some degree? Have we made assumptions that disenfranchised some portion of our students? Have we privileged students who were eager to share and whom we thought we could count on to have something useful to say? Who were the students on the periphery of our classrooms, the ones we thought of as "difficult" or "confused" or "lazy"? Were whole-group discussions dominated by a handful of students? Who were they? Was there a division by gender, race, ethnicity, or language in terms of who spoke, whose representations were considered, whose ideas were repeated or recorded? What happened when a student spoke hesitantly, slowly, backtracked, or broke off? Our own hesitance to "embarrass" a student, our own shyness, our own uncertainties, agendas, time pressures—how have these shaped our practice?

It is not always easy to see, for ourselves, the ways in which we are systematically privileging some students and sidelining others. Pairing with a trusted colleague or two to identify goals and devise an action plan related to a commitment to equitable participation can provide a structure for being responsible to that commitment. The pair might observe each other's lessons, take notes about which students participate and in what ways, and note the nature of the questions posed to different students and the teacher's responses to students' contributions.

In the six classrooms in this book, the teachers made concerted efforts throughout the year to develop a classroom community in which everyone had an opportunity to voice their thinking, build a network of mathematical concepts, and consolidate their nascent understanding. The teachers worked with their students to establish classroom norms that would provide everyone a chance to speak and establish a spirit of collaboration. They carefully selected whose work to present to the whole group, considering both the mathematical content to be explored and the distribution of students whose thinking would be affirmed. They took advantage of the variety of representations created by students as a vehicle for deeper understanding on the part of everyone in the class. And, as the year went on, they reflected, reconsidered, revised, and tried new approaches. As they put these practices into play, students became initiators and advocates for their own learning.

Final Thoughts

As you come to the end of this book, we hope that you see it not as a closing but an opening. We hope that some of the suggestions in this chapter, as well as the main ideas and take-aways from each of the previous chapters, provide doorways to your ongoing reflection and action. We urge you to seek out ongoing, collaborative, focused learning—about mathematics, about student thinking, and about equitable participation. And we wish for you the great pleasure of remaining curious, finding joy and beauty in making new connections, and building a web of understanding, commitment, and determination that grows larger and denser.

Taking a Next Step

Based on your reading and reflections, what do you want to work on in your own practice? Think about what you want to accomplish in the long term, medium term, and short term. What needs to happen that would allow you to reflect further and then put your reflections into action? Are there opportunities to work with a colleague or colleagues toward your goals? Whatever you choose for your own continued learning, how does it embrace the dual commitment to deep and rigorous mathematics and to equitable participation for every student?

Appendix A

Mathematical Generalizations Explored in This Book

Most of the chapters in this book explore one or two generalizations about the operations. Below we have listed those generalizations and the chapters in which they appear. We've included symbolic notation of the generalizations for interested readers, but we don't necessarily recommend introducing such notation to elementary students. Similarly, we've identified the formal name of the generalizations that are conventional Laws of Arithmetic, though elementary students are not expected to learn those terms.

1. Changing the order of addends in an addition expression does not change the sum (commutative law of addition).

   Given numbers a and b, $a + b = b + a$.

   See Introduction

2. Changing the order of factors in a multiplication expression does not change the product (commutative law of multiplication).

   Given numbers a and b, $a \times b = b \times a$.

   See Chapter 3, Introduction to Part Two, Chapter 6

3. Given two addends and their sum, the difference between the sum and one of the addends is the other addend.

   Given numbers a, b, and c, if $a + b = c$, then $c - a = b$ and $c - b = a$.

   See Chapter 1, Chapter 3, Introduction to Part Three, Chapter 8, Chapter 12

4. Given two non-zero factors and their product, if the product is divided by one of the factors, the quotient is the other factor.

   Given numbers a, b, and c not equal to 0, if $a \times b = c$, then $c \div a = b$ and $c \div b = a$.

   See Chapter 3, Introduction to Part Three

5. Given an addition expression, if you add 1 (or some amount) to one addend and subtract 1 (or the same amount) from another addend, the sum remains the same.

   Given numbers a, b, and n, $a + b = (a + n) + (b - n)$

   See Chapter 4, Chapter 5, Chapter 6, Chapter 8, Chapter 9

6. Given a subtraction expression, if you increase both numbers by 1 (or by the same amount), the difference remains the same.

   Given numbers a, b, and n, $a - b = (a + n) - (b + n)$

   See Chapter 5

7. Given a multiplication expression, if you multiply one factor by 2 and divide the other factor by 2, the product remains the same.

   Given numbers a and b, $a \times b = (a \times 2) \times (b \div 2)$

   See Chapter 11

8. Given an addition expression, if 1 (or some amount) is added to an addend, the sum increases by 1 (or that amount) (associative law of addition).

   Given numbers a, b, and c, $a + (b + c) = (a + b) + c$.

   See Chapter 6, Chapter 9, Chapter 10, Chapter 11

9. Given a subtraction expression, if 1 is added to the starting amount (minuend), the difference increases by 1.

   Given numbers a and b, $(a + 1) - b = (a - b) + 1$.

   See Chapter 6

10. Given a subtraction expression, if 1 is added to the amount subtracted (subtrahend), the difference decreases by 1.

    Given numbers a and b, $a - (b + 1) = (a - b) - 1$.

    See Chapter 6, Chapter 7

11. Given a multiplication expression, if you add 1 to a factor, the product increases by the other factor (a special case of the distributive law of multiplication over addition).

    Given numbers a and b, a $\times (b + 1) = ab + a$ and $(a + 1) \times b = ab + b$.

    See Chapter 7, Chapter 10, Chapter 12

Appendix B
Resources for Continued Learning

1. Continued Learning About Equitable Participation in the Elementary Mathematics Classroom.

(A) Aguirre, J., Mayfield-Ingram, K., & Martin, D. B. (2024). *The impact of identity in K–12 mathematics: Rethinking equity-based practices.* (Expanded edition.) National Council of Teachers of Mathematics.

The first edition of this book, published in 2013, influenced our thinking about the intersection of rigorous mathematics and equitable participation during the project on which this book is based. In particular, it introduced us to the idea of *collective mathematical agency*, which we have found so helpful for our work. The authors focus on students' multiple identities and on how teachers can draw on these identities, illustrating their ideas with classroom vignettes. They focus on five equity-based mathematics teaching practices:

- Going deep with mathematics
- Leveraging multiple mathematical competencies
- Affirming mathematics learners' identities
- Challenging spaces of marginality
- Drawing on multiple resources of knowledge

(B) Chval, K. B., Smith, E., Trigos-Carillo, L., & Pinnow, R. J. (2021). *Teaching math to multilingual students: Positioning English learners for success.* Corwin.

While visiting the classrooms of our Collaborating Teachers, we recognized how their practices supported the voices of multilingual learners, and from time to time our Critical Friends commented on such strategies. For a more in-depth treatment of equitable participation for multilingual learners in mathematics, we recommend the book by Katherine Chval and colleagues. It focuses on strength-based approaches for supporting multilingual students' learning, developing these students' leadership roles in the classroom, and strategies for connecting with students' families.

(C) The Forum for Equity in Elementary Mathematics (https://www.terc.edu/mathequityforum/) offers a number of resources to support teachers who are focused on interweaving deep mathematics and equitable participation. These include

1. *Teacher Reflection Tools* designed to help teachers keep equitable participation at the forefront as they plan for and facilitate whole-group discussions and small group/pair work.

2. *Student Reflection Tool* designed to support teachers in making student reflection a regular part of their practice. It provides a bank of questions and sample questionnaires to use with students as well as a planning tool that guides teachers in identifying goals, choosing questions, analyzing student responses, and considering next steps, based on what teachers are learning from students' responses

3. *Curriculum Excursions*—sequences of related classroom activities that encourage students to apply the mathematics they are learning to the contexts of their own and others' lives and communities.

The Forum is a project of TERC.

2. Continued Learning About Noticing and Conjecturing About the Operations

The class sessions in this book show students engaged in noticing, conjecturing, and representing generalizations about the operations (one facet of early algebra). Here are two books we developed, in collaboration with groups of teachers, that focus on this math content in the elementary classroom.

(A) Russell, S. J., Schifter, D., Kasman, R., Bastable, V., & Higgins, T. (2017). *But why does it work? Mathematical argument in the elementary classroom.* Heinemann.

This book, designed for individuals as well as study groups, offers readers a teaching model for integrating mathematical argument into instruction. The teaching model includes five phases: noticing relationships across sets of problems, equations, or expressions; articulating a claim about what has been noticed; investigating the claim through representations; using representations to demonstrate why the claim must be true or not; and extending thinking from one operation to another. Each phase is illustrated by written and video cases from collaborating teachers' classrooms.

The book includes eight instructional sequences designed for Grades 1–5. Each instructional sequence is a series of about 20 to 25 short sessions that focus on two or three related generalizations. Teachers implement a lesson sequence over the course of a semester in the way that they might use Number Talks or Classroom Routines.

(B)   Russell, S. J., Schifter, D., & Bastable, V. (2011). *Connecting arithmetic to algebra*. Heinemann.

Russell, S. J., Schifter, D., & Bastable, V. (2012). *Connecting arithmetic to algebra, course facilitator's guide.* Heinemann.

This book lays out the territory of engaging young students in mathematical argument. It focuses on algebra readiness as the development of ways of thinking that underlie both arithmetic and algebra. Illustrated with rich examples from teachers' own classroom accounts, it includes chapters on how the range of learners in the classroom participate in early algebra, on the role of symbolic notation in the early grades, and on looking ahead to the middle grades. An online Facilitator's Guide provides a structure and resources for using this book as the basis for a professional development course or study group.

3. Continued Learning About Elementary Mathematics Content and How Students Think About Mathematics

The *Developing Mathematical Ideas* (DMI) series is a professional development curriculum designed to help teachers think through the major ideas of K–8 mathematics and examine how children develop those ideas. At the heart of the materials are sets of classroom episodes (cases) illustrating student thinking as described by their teachers. In addition to case discussions, the curriculum offers teachers opportunities: to explore mathematics in lessons led by facilitators; to share and discuss the work of their own students; to view and discuss video clips of mathematics classrooms; to write their own classroom cases; and to read overviews of related research.

DMI, published by the National Council of Teachers of Mathematics, consists of seven modules, each designed for eight 3-hour sessions. Primary authors of the series are Deborah Schifter, Virginia Bastable, and Susan Jo Russell.

- **Number and Operations Part 1: Building a System of Tens.** Participants explore the base-ten structure of the number system, consider how that structure is exploited in multidigit computational procedures, and examine how basic concepts of whole numbers reappear when working with decimals. (See https://www.nctm.org/Store/Products/Number-and-Operations,-Part-1--Building-A-System-of-Tens-Facilitators-Package/)

- **Number and Operations Part 2: Making Meaning for Operations.** Participants examine the actions and situations modeled by the four basic operations. The seminar begins with a view of young children's counting strategies as they encounter word problems, moves to an examination of the four basic operations on whole numbers, and revisits the operations in the context of rational numbers. (See https://www.nctm.org/Store/Products/Number-and-Operations,-Part-2--Making-Meaning-for-Operations-Facilitators-Package/)

- **Number and Operations Part 3: Reasoning Algebraically About Operations.** Participants examine generalizations at the heart of the study of operations in the elementary grades. They express these generalizations in common language and in algebraic notation, develop arguments based on representations of the operations, study what it means to prove a generalization, and extend their generalizations and arguments when the domain under consideration expands from whole numbers to integers. (See https://www.nctm.org/Store/Products/Number-and-Operations,-Part-3--Reasoning-Algebraically-About-Operations-Facilitators-Package/)

- **Algebra: Patterns, Functions, and Change.** Participants discover how the study of repeating patterns and number sequences can lead to ideas of functions, learn how to read tables and graphs to interpret phenomena of change, and use algebraic notation to write function rules. While its particular emphasis is on linear functions, the seminar also provides opportunities to explore quadratic and exponential functions and to examine how various features of a function are seen in graphs, tables, or rules. (See https://www.nctm.org/Store/Products/Algebra--Patterns,-Functions,-and-Change-Facilitators-Package/)

- **Geometry: Examining Features of Shape.** Participants examine aspects of 2D and 3D shapes, develop geometric vocabulary, and explore both definitions and properties of geometric objects. The seminar includes a study of angle, similarity, congruence, and the relationships between 3D objects and their 2D representations. (See https://www.nctm.org/Store/

Products/Geometry--Examining-Features-of-Shape-Facilitators-Package/)

- **Geometry: Measuring Space in One, Two, and Three Dimensions.**
  Participants examine different attributes of size, develop facility in
  composing and decomposing shapes, and apply these skills to make sense
  of formulas for area and volume. They also explore conceptual issues of
  length, area, and volume, as well as their complex interrelationships.
  (See https://www.nctm.org/Store/Products/Geometry--Measuring-Space
  -in-One,-Two,-and-Three-Dimensions-Facilitators-Package/)

- **Statistics: Modeling With Data.** Participants work with the collection,
  representation, description, and interpretation of data. They learn what
  various graphs and statistical measures show about features of the data,
  study how to summarize data when comparing groups, and consider
  whether the data provide insight into the questions that led to data
  collection. (See https://www.nctm.org/Store/Products/Statistics
  --Modeling-with-Data-Facilitators-Package/)

References

Aguirre, J., Anhalt, C., Cortez, R., Turner, E., & Simic-Muller, K. (2019). Engaging teachers in the powerful combination of mathematical modeling and social justice: The Flint water task. *Journal of Mathematics Teacher Education, 7,* 7–21. http://works.bepress.com/julia _aguirre/25/

Aguirre, J., Mayfield-Ingram, K., & Martin, D. B. (2024). *The impact of identity in K–12 mathematics: Rethinking equity-based practices* (Expanded ed.). National Council of Teachers of Mathematics.

Cirillo, M., Bartell, T. G., & Wager, A. A. (2016). Teaching mathematics for social justice through mathematical modeling. In C. Hirsch (Ed.), *Mathematical modeling and modeling mathematics* (pp. 87–96). National Council of Teachers of Mathematics.

Delpit, L. (2012). *"Multiplication is for white people": Raising expectations for other people's children.* New Press.

Duckworth, E. (1996). *"The having of wonderful ideas" and other essays on teaching and learning.* Teachers College Press.

Hardy, G. H. (1967). *A mathematician's apology.* Cambridge University Press.

Joseph, N. M., & Alston, N. V. (2018). We fear no number: Humanizing mathematics teaching and learning for Black girls. In I. Goffney, R. Gutiérrez, & M. Boston (Eds.), *Rehumanizing mathematics for Black, Indigenous, and Latinx students* (pp. 51–62). National Council of Teachers of Mathematics.

Ladson-Billings, G. (2006). From the achievement gap to the education debt: Understanding achievement in U.S. schools. *Educational Researcher, 35*(7), 3–12.

Ladson-Billings, G. (2007). Pushing past the achievement gap: An essay on the language of deficit. *The Journal of Negro Education, 76*(3), 316–323.

Leonard, J., & Martin, D. B. (2013). *The brilliance of Black children in mathematics: Beyond the numbers and toward new discourse.* Information Age.

Lewis, K. (2018). Difference, not deficit: Reconceptualizing mathematical learning disabilities. *Journal of Education, 196*(2), 39–57. https://doi .org/10.1177/002205741619600203

National Council of Supervisors of Mathematics & TODOS. (2016). *Mathematics education through the lens of social justice: Acknowledgement, actions, and accountability.* https://www .mathedleadership.org/position-papers/

National Council of Supervisors of Mathematics & TODOS. (2021). *Positioning multilingual learners for success in mathematics.* https://www .mathedleadership.org/position-papers/

National Council of Teachers of Mathematics. (2000). *Principles and standards for school mathematics.* Author.

National Governors Association Center for Best Practices & Council of Chief State School Officers. (2010). *Common core state standards for mathematics.* Author.

Paley, V. G. (1986). On listening to what the children say. *Harvard Educational Review, 56*(2), 122–131.

Russell, S. J., Schifter, D., Kasman, R., Bastable, V., & Higgins, T. (2017). *But why does it work? Mathematical argument in the elementary classroom.* Heinemann.

Steen, L. A. (1990). *On the shoulders of giants.* National Research Council.

Zwiers, J., & Crawford, M. (2011). *Academic conversations: Classroom talk that fosters critical thinking and content understandings.* Routledge.

Index

A Sage Company

Helping educators make the greatest impact

CORWIN HAS ONE MISSION: to enhance education through intentional professional learning.

We build long-term relationships with our authors, educators, clients, and associations who partner with us to develop and continuously improve the best evidence-based practices that establish and support lifelong learning.

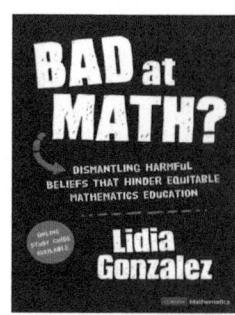

**JENNIFER M. BAY-WILLIAMS,
JOHN J. SANGIOVANNI, ROSALBA SERRANO,
SHERRI MARTINIE, JENNIFER SUH,
C. DAVID WALTERS, SUSIE KATT**

Because fluency is so much more than
basic facts and algorithms.

Grades K–8

**MARIA DEL ROSARIO
ZAVALA,
JULIA MARIA AGUIRRE**

Discover innovative equity-
based culturally responsive
mathematics instruction that
unlocks the mathematical
heart of each student.

Grades K–8

**IVANNIA SOTO,
THEODORE RUIZ SAGUN,
MICHAEL BEIERSDORF**

Focus on the literacy
opportunities that multilingual
students can achieve when
language scaffolds are
taught alongside rigorous
math and science content.

Grades K–8

**JOHN J. SANGIOVANNI, SUSIE KATT,
LATRENDA D. KNIGHTEN, GEORGINA RIVERA,
FREDERICK L. DILLON, AYANNA D. PERRY,
ANDREA CHENG, JENNIFER OUTZS, KAREN MESMER,
ENYA GRANDOS, KEVIN GANT, LAURA SHAFER**

Actionable answers to your most pressing
questions about teaching elementary math,
secondary math, and secondary science.

Elementary, Secondary

**CHRISTA JACKSON, KRISTIN L. COOK,
SARAH B. BUSH,
MARGARET MOHR-SCHROEDER,
CATHRINE MAIORCA, THOMAS ROBERTS**

Help educators create integrated STEM
learning experiences that are inclusive for all
students and allow them to experience STEM
as scientists, innovators, mathematicians,
creators, engineers, and technology experts!

Grades PreK–5 and Grades 6–12